Return On Quality

Measuring
the Financial Impact
of Your Company's
Quest for Quality

Roland T. Rust

Anthony J. Zahorik

Timothy L. Keiningham

PROBUS PUBLISHING COMPANY
Chicago, Illinois
Cambridge, England

ISBN 1-55738-547-5

Printed in the United States of America

BB

1 2 3 4 5 6 7 8 9 0

To the memory of
Bruce D. Henderson,
a thought leader who found opportunities
where others saw only problems

April 30, 1915–July 20, 1992

CONTENTS

LIST OF FIGURES AND TABLES

FOREWORD

How many of us can actually measure the return on quality? Is there really a link between customer satisfaction and market share? What is the link and how can you establish it in your business? To establish the link in your business you need to know exactly what your customers require for what they are willing to pay. A good customer satisfaction measurement program will give you the strategic intelligence you need. But just what do you have to do to get valid customer satisfaction data? How do you analyze the data to convert it into useful, actionable information? What are the proven ways to tie the information into process improvement programs and reengineering efforts? How can you integrate the information with effective strategic plans? If you can answer these questions, you know the linkage between customer satisfaction and marketplace success. If you can answer these questions, you can establish the linkage for your business and develop a strong, sustainable, competitive advantage.

The measure of success of a business is the value it creates for its owners. But the way to achieve success is to satisfy customers with quality products and services at a price that represents the best value in the market. Customers really have all the votes. Because of this I like to say business organizations can only be successful if they (1) find the customers' needs, and (2) fulfill them better than any competitor can.

Studying customer satisfaction and dissatisfaction over the years, I have found that no one person in any organization can totally satisfy a customer. But any one person can totally *dissatisfy* a customer. Completely satisfying a customer requires totally aligned and focused people. Someone once said that organizations exist to achieve together what cannot be accomplished alone.

For all the talk and bravado in the literature and corporate speeches today about customer satisfaction, there is a void on useful and practical ways to capture customer needs, determine how well you are satisfying those needs, and build actionable plans and measurements to help win in the marketplace. Customer satisfaction in most companies is more of a slogan than a practical science and art. This book provides the information and tools to fill that void. It provides, I believe, the knowledge leaders of small and large businesses need to increase their chances of success. It explains in simple terms what works, what doesn't work, and why. It tells the reader not only what to do, but how to do it. In some respects it is profound. In other respects it is common sense. Albert Einstein once said, "The whole of science is nothing more than the refinement of everyday thinking." This book provides the everyday thinking business leaders need to win customer votes in the marketplace.

Raymond E. Kordupleski
Customer Satisfaction Director
AT&T

The importance of product and service quality has been discussed and analyzed at length during the past decade. As a result, business leaders today are more focused on achieving benchmarks of quality and are implementing standards of measurement to help achieve these goals within their companies.

While most business people agree that quality is important to the success of a company, very few are able to discuss the specific cost of quality and its financial impact on their business. As this book illustrates, quality must be viewed as more than a concept. By listening to customers and defining quality in their terms, companies can increase the overall level of customer retention and substantially impact profit margins.

The practical principles the authors detail in *Return on Quality* will assist progressive managers in quantifying the value of a quality strategy in their organizations.

Mark C. Wells
Senior Vice President
Marketing
Promus Hotel Group

PREFACE

The 1980s marked the beginning of a quality revolution for U.S. corporations. Hundreds of books were written espousing the importance of quality management and providing stories of firms that had achieved financial success through quality improvement. The U.S. government even began promoting the importance of quality with its founding of the Malcolm Baldrige National Quality Award.

As a result, virtually every large American corporation has initiated some form of quality-improvement program. Most of these programs, however, are not achieving significant results. Even worse, the failure of several acclaimed, quality-oriented firms has demonstrated that quality is not a guarantee of profits.

This does not mean that quality is not necessary to achieve long-term profitability. In fact, there is scientific evidence of a relationship between quality and profits. Instead, it suggests that businesses are not effectively evaluating and implementing their quality-improvement programs.

Currently, companies embark on quality initiatives without any idea of what the likely bottom-line impact will be. As a result, they have no way to determine which actions are most important or to predict if their efforts will actually benefit their firms. Thus, they are literally taking a journey without any map to guide them to their destination. What is needed is a way for firms to assess the likely profit impact of various quality-improvement alternatives.

Thanks to the work of many different researchers, most of the causal links between company actions and customer reactions as they pertain to quality improvement are well understood. However, no one has pulled this information together into a comprehensive quality impact measurement system. The purpose of this book is to show managers how to tie this information together so that they can measure their Return on Quality (ROQ).

ACKNOWLEDGMENTS

It is impossible to thank all of the people who provided support in the writing of this book. There are, however, several firms and individuals whose contributions were invaluable to its completion.

We are indebted to the partner companies of Vanderbilt's Center for Services Marketing: AT&T, Northern Telecom, NationsBank, Union Planters National Bank, the Promus Companies, and Walker: Customer Satisfaction Measurement. They provided advice and encouragement for the development of both the ROQ model and the manuscript. We would especially like to thank Union Planters National Bank and Promus Companies for acting as beta test sites for the ROQ model. Particular acknowledgment must be given to Mark Wells, Bala Subramanian, David Williams, Gary Howlett, Ed Routon, Jim Alcott, and Jennifer Knickerbocker, who were instrumental in implementing ROQ.

Significant amounts of research were necessary to this undertaking. We wish to thank Patrali Chatterjee for helping with much of the library research, thereby allowing us to complete this project on time. Also, Michiko Keiningham designed many of the exhibits used in this book.

The actual writing benefited greatly from the contributions of several individuals. Dr. Germain Boer provided invaluable insight and comments regarding the Cost of Quality analysis discussed in Chapter 7. N. Laddie Cook, James A. Welch, and Dr. P. Ranganath Nayak of Arthur D. Little, Inc. showed how ROQ fit into the High-Performance Business framework and coauthored Chapter 10 of the book with us. Richard Royce provided review and suggestions regarding several chapters. Also, Raymond Kordupleski of AT&T, Marshall Weems of Financial Selling Systems, and Donald Jackson of the Jackson Consulting Group contributed greatly to our knowledge of quality-improvement implementation.

Our deepest thanks go to our loved ones for accepting the sacrifices required and putting up with us (for the most part) while we were working long hours to complete this project.

QUALITY: WHAT WENT WRONG?

QUALITY AND RESULTS: WHAT HAPPENED?

There seems to be plenty of evidence that high quality and profitability go together. Beginning at least as early as Peters' and Waterman's book *In Search of Excellence* in 1982,[1] a steady stream of books has issued from the business press describing firms for which quality programs have led to new levels of financial success. Trade magazines regularly publish special issues describing the successes achieved by industry leaders in quality.

But stories aren't proof. While 93 percent of large U.S. corporations have some form of quality-improvement program, several recent studies suggest that many of these programs are failing.[2] A McKinsey & Company study found that of quality programs in place for more than two years, two-thirds have failed due to poor results.[3] Also, separate surveys by Arthur D. Little and A. T. Kearney revealed that only one-third of companies with total quality programs have achieved significant results.[4]

Trouble with quality programs is not limited to inexperienced firms, either. Several acclaimed quality-oriented companies have experienced difficulty. The Wallace Company, Florida Power & Light, Centennial Medical Center, and IBM provide excellent examples of such firms.

Wallace Company

In 1989, Wallace Company, a Houston pipe and valve distributor, attempted to address a host of quality issues, including 72 different problems just with delivery and invoicing, by purchasing new computers, redesigning truck-loading systems, and instituting other measures.[5] The company was recognized for its efforts by becoming the first small business to win the Malcolm Baldrige National Quality Award in 1990.

However, being honored for quality operations did not translate into profits. Although market share rose from 10.4 percent to 18 percent, overhead also went up by $2 million per year. Because of the increased overhead, Wallace was forced to raise prices, which customers weren't always willing to pay.[6] As a result, at the same time that the company was receiving the award, it was losing $300,000 a month.[7] By January 1992 Wallace was forced to declare Chapter 11 bankruptcy. By August of that year, the company was out of business.[8]

Florida Power & Light

Florida Power & Light (FP&L) began a companywide quality-improvement program in 1981. By June of 1986, the firm was recognized as a quality leader by its peers in the utility industry when it was presented with the Edison Award.[9] In 1989, FP&L received even greater recognition when it became the only U.S. company to win Japan's prestigious Deming Prize for quality. Winning these awards made FP&L a symbol of American success at adapting Japanese quality-improvement methods.

This success, however, did not come cheap. FP&L spent $2.85 million in pursuit of the Deming Prize.[10] However, the company found itself with morale problems and financial difficulty less than a year after winning.[11] Further, the Florida Public Service Commission refused to allow FP&L to pass along all of the costs it incurred pursuing the Deming Prize to rate-payers.[12] In fact, the prize was losing its luster even before the winner was announced.

While the utility was seeking the Deming Prize, the company's directors were moving FP&L's chairman (and chief quality advocate), John Hudiburg, "upstairs" by making him chairman emeritus.[13] At about the time the company announced that it had won the prize, he retired. Hudiburg left because James Broadhead, the chairman of FP&L's holding company, made it clear that he did not support past management practices.[14] By June of 1990, Broadhead had dismantled many of the procedures considered essential to the company's Total Quality Control program and eliminated the Quality Improvement Department, the Quality Improvement Promotion Group, and Quality Support Services.[15]

Centennial Medical Center

In early 1989, Hospital Corporation of America (HCA) brought in William W. Arnold as president of its flagship operation—Centennial Medical Center, based in Nashville, Tennessee. At the time, he was lauded as a visionary who would successfully implement "person-centered leadership," a management style adapted from the teachings of respected quality expert W. Edwards Deming, to guide the company.

Arnold sought to include every employee in management decisions and worked hard to show employees that he sincerely cared about their opinions and wanted to hear them. Any employee was welcome to come and speak to him about anything at all. As a symbol of his accessibility, Arnold had the door to his office removed from its hinges and placed in the waiting area outside.

Because of Arnold's unique management approach, he became a noted speaker on quality issues. He also coauthored a book, *The Human Touch,* in which he detailed his experiences with Centennial and argued that person-centered leadership leads to increased productivity and profits. A quote from this book reflects the intensity Arnold held for his management philosophy:

> Several department heads and executives have shared with me that they had a simple hope: that it would all end far short of a blistering, blood-red bottom line. I may have been frustrated, but they were scared. I seemed to be paying more attention to associates' needs than to fiscal responsibility. But I knew I had to pay as much attention to my office door as I did to that bottom line.[16]

Unfortunately, the bottom line results did not appear. Although Centennial is the third largest hospital in Tennessee and the HCA flagship, it lost nearly $1 million on revenues of $161 million in 1991.[17] Arnold was fired in May of 1993.

Centennial's new president is reported to be committed to following quality principles, but has made it clear that ". . . you must still be willing to make tough decisions when you can't afford the luxury of time."[18] Also, the door is back on its hinges.

IBM

In early 1990, IBM embarked on a companywide quality program dubbed Market-Driven Quality.[19] One of the key components of the program was the adoption of a "six sigma" approach to eliminating defects. (Six sigma is a statistical term popularized by Motorola indicating that a process produces approximately 3.4 defects per million operations.[20])

IBM chairman John Akers fully embraced the firm's quality-improvement program, stating, "I am almost religious about my dedication to quality improvement. My number-one priority is to have Market-Driven Quality implemented as broadly and as deeply as it can be by the time I retire."[21] He even wrote an article, "World-Class Quality: Nothing Less Will Do," which expressed his enthusiasm for IBM's quality-improvement program. In the article he writes:

> As with many long-established institutions, IBM needed to renew itself to compete successfully in a rapidly changing world. The renewal was aimed at improving our customer relationships, the competitiveness of our products and services, and our efficiency—in short, at achieving 100 percent satisfaction among our customers. We established a working premise that would drive IBM's business strategies from

then on: if we could be world class—the best—at satisfying the needs and wants of customers in the markets we chose to serve, all else would follow, including market share, revenue, and profits."[22]

IBM's quality-improvement efforts have resulted in significant recognition. A division of the company, IBM Rochester, won the Malcolm Baldrige National Quality Award in December of 1990. IBM also received the George M. Low Trophy, NASA's Quality and Excellence Award, in 1992.[23]

So far, however, IBM's Market-Driven Quality program has not resulted in improved financial performance. In fact, IBM has been forced to lay off tens of thousands of workers and record billions in losses. As a result, Akers has been replaced by former RJR-Nabisco CEO Louis Gerstner.

DOES QUALITY LEAD TO PROFITS?

These examples make it clear that although high quality may be necessary to be profitable (or even to compete) in many industries these days, it is not a guarantee of profits. What caused the quality programs at Wallace Company, Florida Power & Light, Centennial Medical Center, and IBM to stumble? While there are circumstances unique to each example, all of these companies failed to link their quality programs to the bottom line through cost reductions and/or revenue increases. As a result, there was no way for them to determine what actions were most important or to predict if their efforts would actually benefit the company.

FIGURE 1-1   PIMS CHART

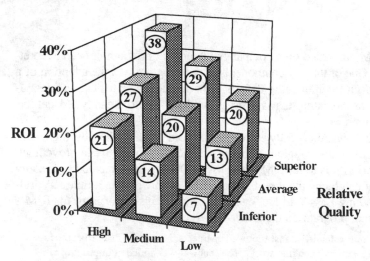

Reprinted with the permission of The Free Press, a Division of Macmillan, Inc. from THE PIMS PRINCIPLES: Linking Strategy to Performance, by Robert Buzzell and Brad Gale. Copyright © 1987 by The Free Press.

In spite of these and other notable exceptions, however, there is some scientific evidence of a relationship between quality and profits. The Strategic Planning Institute in Boston maintains a large database, called the PIMS (Profit Impact of Marketing Strategy) data, containing information on profitability and a large number of strategic variables for companies in a variety of industries. In a series of studies of the PIMS data researchers have found strong positive relationships between return on investment (ROI) and reported quality levels, and between quality and market share growth (see Figure 1-1).[24]

Where do the higher profits come from? Figure 1-2 shows the general effect of quality on the costs and sales of a firm. The main effects of quality on profits

FIGURE 1-2   HOW QUALITY LEADS TO PROFITS

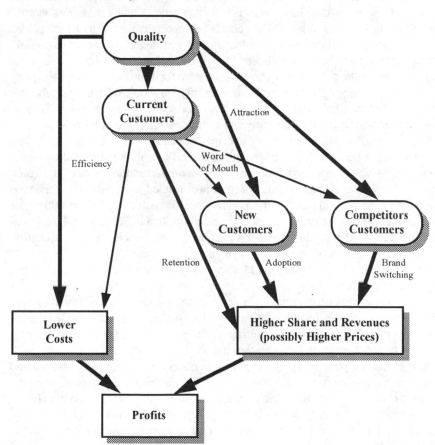

are realized through lower costs due to efficiencies achieved, higher customer retention, greater attraction of new customers, and the potential to charge higher prices.

These effects aren't necessarily the same for all firms and industries. Studies of the PIMS data found that the relative importance of each of these factors varied widely across different industries.[25] For example, in some industries the high-quality firms fill high-priced niche markets; in others, the demand for high-quality products can be the basis for a high-volume, low-price market domination strategy due to the lower costs caused by efficiencies and experience curve effects.

However, it appears that not many companies carefully track the source of profits from their quality programs, and fewer try to predict the effects before spending. The U.S. General Accounting Office found that only five of 22 finalists for the Baldrige Award—firms that must be considered among the country's elite quality-oriented firms—had measured the cost savings from their quality programs. There are several reasons for this. First, there is a feeling among many managers that the value of quality is unquantifiable. The general perception is that many of the effects are hard to measure and require very subjective cost estimates. Second, many quality-oriented managers simply don't believe that quality should be subject to financial criteria. For example, a director of quality assurance for Motorola's computer group was quoted as saying, "I have a problem with anyone who asks for a cost/benefit analysis. . . . We don't allow defects to be characterized as minor or major. They all must be eliminated [because] we simply can't afford to have poor quality."

In a survey of electronics industry executives, profits were rated a distant third as an anticipated benefit of quality programs behind product quality and customer satisfaction.[26] Even W. Edwards Deming, who is considered the father of the quality revolution, sees no value in financial measures related to quality. His feeling is that many of the financial benefits from quality are "invisible and unknowable" due to higher-order effects of improved morale and multiplied efficiency, and that measurable effects will include only a trivial part of the process.[27] This question remains an area of open debate in quality management circles.

However, spending on quality is like any other resource allocation decision; it is expected to produce returns that are greater than the costs. This is true even for firms that do not measure the financial impact of their quality improvement efforts, since the primary reason for the popularity of the quality movement is the implied link between quality and profits. Ultimately, the justification for any quality program must be its connection to the bottom line.

As a result, the pressure is on companies to manage their limited resources and to direct their spending to where it counts most. It is financially impossible for most firms to undertake all possible quality-improvement efforts. Further, the benefits of different programs can vary dramatically for different companies. Therefore, managers need to determine the return on potential quality-improvement efforts in order to maximize their investments.

Unfortunately, managers have had a difficult time estimating the expected returns from their quality expenditures. Usually, they have been forced to accept on faith that their investments in quality will ultimately benefit the company. So how is a manager to decide which areas warrant spending, and how much should be spent on them? Thanks to the work of many different researchers, most of the causal links between product quality and customer retention are well understood.[28] Research has also shown that it is possible to statistically identify links between individual product attributes and customer behavior.[29] However, while the individual parts of the relationship between corporate actions and customer reactions have been heavily researched, no comprehensive quality impact measurement system exists. These pieces need to be assembled into a coherent model of the entire process, from quality to profits. The purpose of this book is to show managers how to pull these pieces together so that they can measure their Return on Quality (ROQ).

Figure 1-3 shows the steps involved in determining a company's ROQ. The steps can be broken down into four main sections: (1) Exploratory Research, (2) Quantitative Research, (3) Impact of Quality on Satisfaction, and (4) Market Share and Profit Impact. The components of these steps will be discussed in later chapters in the book.

The central chain of events that leads from quality to profits can be summarized as follows:

The first thing to note about the above chain of events is that it focuses on customer retention and does not include three sources of profits generated by quality improvement: cost reductions due to increased efficiency, the attraction of new customers resulting from positive word-of-mouth, and the ability to charge higher prices. Cost reductions resulting from quality improvement can and should be included when determining a company's ROQ. Measuring this effect, however, is not contingent upon the links in the above chain. Rather, this effect can be measured separately and included when determining the profit impact. We will discuss how to measure a firm's "cost of quality" in order to determine what, if any, cost savings can be achieved as a result of quality improvement.

FIGURE 1-3   STEPS IN DETERMINING RETURN ON QUALITY (ROQ)

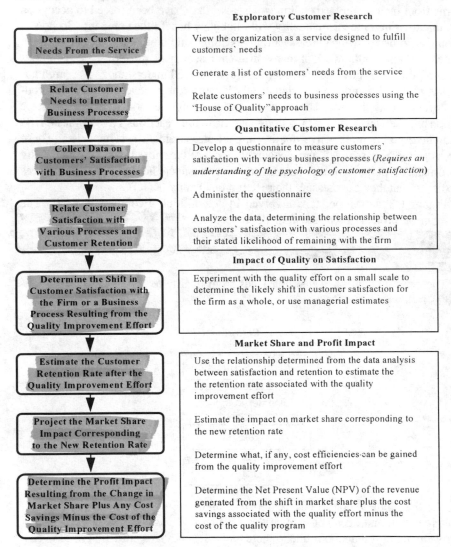

Exploratory Customer Research

Determine Customer Needs From the Service
- View the organization as a service designed to fulfill customers' needs
- Generate a list of customers' needs from the service

Relate Customer Needs to Internal Business Processes
- Relate customers' needs to business processes using the "House of Quality" approach

Quantitative Customer Research

Collect Data on Customers' Satisfaction with Business Processes
- Develop a questionnaire to measure customers' satisfaction with various business processes (*Requires an understanding of the psychology of customer satisfaction*)
- Administer the questionnaire

Relate Customer Satisfaction with Various Processes and Customer Retention
- Analyze the data, determining the relationship between customers' satisfaction with various processes and their stated likelihood of remaining with the firm

Impact of Quality on Satisfaction

Determine the Shift in Customer Satisfaction with the Firm or a Business Process Resulting from the Quality Improvement Effort
- Experiment with the quality effort on a small scale to determine the likely shift in customer satisfaction for the firm as a whole, or use managerial estimates

Market Share and Profit Impact

Estimate the Customer Retention Rate after the Quality Improvement Effort
- Use the relationship determined from the data analysis between satisfaction and retention to estimate the the retention rate associated with the quality improvement effort

Project the Market Share Impact Corresponding to the New Retention Rate
- Estimate the impact on market share corresponding to the new retention rate
- Determine what, if any, cost efficiencies can be gained from the quality improvement effort

Determine the Profit Impact Resulting from the Change in Market Share Plus Any Cost Savings Minus the Cost of the Quality Improvement Effort
- Determine the Net Present Value (NPV) of the revenue generated from the shift in market share plus the cost savings associated with the quality effort minus the cost of the quality program

The attraction of new customers (above the current rate) is not included in the measurement of ROQ, because this effect occurs over longer periods of time and is difficult to measure. Also, since not all businesses can charge higher prices because of increased quality, this stage is not included in the ROQ analysis. As Figure 1-3 shows, measuring the impact of quality on profits requires that firms understand what attributes customers use to evaluate a service and what factors drive customer satisfaction. Further, managers need some way to collect information that will allow them to make valid inferences concerning the relationship between customers' satisfaction with various aspects of an organization's products or service and customer retention. The market share and profit impact can then be derived for various quality-improvement options.

EVERYTHING IS A SERVICE

It is important to note that the above chain of events lists "service performance" as the first link in the chain. The choice of words is not arbitrary. Generally, people think of business as being divided up into a product sector and a service sector. Product sector companies are those whose primary mission is to sell physical products, such as cars, computers, and laundry detergent; service sector companies, on the other hand, are those that sell intangible offerings, such as airline travel, education, and telephone service.

Our conception of companies is different. We see the primary purpose of every business and, in fact, every organization as performing a service for customers. The service always centers upon meeting customer needs. Thus, a mission of management is to identify and fulfill customer needs. To do so, companies must find out what the customer needs are, whether they are being met, and how to meet them better.

Thinking of a company as a service designed specifically to fulfill customer needs is not the natural way for most managers to view their organizations. It is essential, however, if they are to understand how customers perceive their firms. As a result, managers need to understand what is meant by "service performance" so that they can determine the actual benefits that their companies offer to customers.

COMPONENTS OF SERVICE

Service may be broken down into four main components: physical product, service product, service environment, and service delivery (see Figure 1-4).[30] The *physical product* is whatever the organization transfers to the customer that can be touched; it is tangible and physically real. Examples include automobiles, computers, books, telephones, houses, televisions, and furniture.

The *service product* is the part of the service that can be planned and designed but is not physical and tangible. For example, a baseball park may decide to have an organist playing, ushers showing people to their seats, and an announcer giving the lineups and saying who is coming up to bat. All of these things must be planned, and they help determine the nature of the service experience.

The *service environment* is the physical backdrop that surrounds the service. For example, going to see a movie is more fun if the theater is clean, has comfortable seats, and has a spacious, well-lighted parking lot. Even though the customer

FIGURE 1-4   THE COMPONENTS OF SERVICE

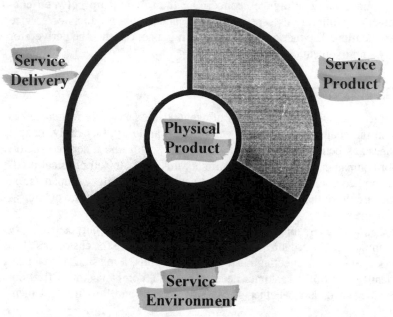

Source: Roland T. Rust and Richard L. Oliver, "Service Quality: Insights and Managerial Implications from the Frontier," in *Service Quality: New Directions in Theory and Practice,* Roland T. Rust and Richard L. Oliver, eds., p. 11, copyright © 1993 by Sage Publications, Inc. Reprinted by permission of Sage Publications, Inc.

doesn't take any of that home, it has an important impact on the service experience.

The *service delivery* is how the service is actually provided. If the service product defines how the service works in theory, the service delivery is how the service works in actual practice. For example, the service design may be that a fast-food customer is greeted within 10 seconds, but the actual service delivery may be hindered by the counter employee being in the back of the store for five minutes, joking with his fellow employees.

Figure 1-5 shows examples of the physical product, service product, service environment, and service delivery for several large industries.

THE PRODUCT SECTOR AND THE SERVICE SECTOR

Traditionally, people have referred to a business as being in the product sector if its purpose was to sell a physical product, and otherwise to consider that the business was in the service sector. The implication of this distinction, from the viewpoint of

FIGURE 1-5   COMPONENTS OF SERVICE: INDUSTRY EXAMPLES

Industry	Physical Product	Service Product	Service Environment	Service Delivery
Auto	the car	pricing	showroom	test drive and sales pitch
		warranty	grounds	repair time
		loans	car lot	negotiation
Hotel	room supplies	pricing	the room	front desk performance
	food	shuttle	pool	room cleaning
		wakeup calls	lobby	promptness of room service
University	diploma	majors	classrooms	teaching
		residence	dormitories	janitorial
		placement	sports fields	
Retail	goods	pricing	sales floor	knowledgeability
		credit		friendliness
		inventory		speed

Figure 1-4, is that product businesses are primarily concerned with the physical product, while service businesses are primarily concerned with service product, service environment, and service delivery. Looking at Figure 1-4, one can say that a product business *should* be concerned with *all* of the service components. In fact, the service elements of service product, service environment, and service delivery are common across *both* the product sector and service sector. Thus we will make no attempt to restrict your attention to the service sector.

Conceptualizing the business as a service, improving service quality, and making sure that customer needs are central to the management of the business are equally valid in both the product and service sectors.

Nevertheless, while service quality is important in the product sector, it is even more important in the service sector. Thus it is enlightening to note the shift toward the service sector in the last 100 years. Figure 1-6 shows dramatically how service employment has risen in the United States from just 30 percent in 1900 to an expected 77 percent in 1995.[31] The implications of this shift are staggering. Focusing on service quality is now critical to the efficiency of the economy of the United States. A similar shift is occurring in all developed or developing nations. This shift toward the service sector also makes it abundantly clear that tomorrow's managers

FIGURE 1-6   THE SERVICE SECTOR: SHARE OF U.S. EMPLOYMENT

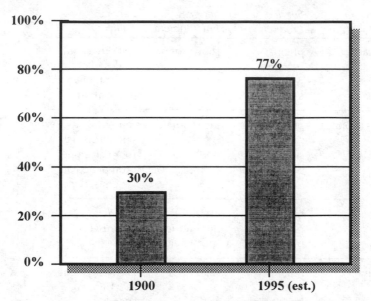

Source: John E. G. Bateson (1989), Managing Services Marketing, 2d ed., The Dryden Press, and Stephen M. Shugan, "Explanations for the Growth of Services," in *Service Quality: New Directions in Theory and Practice,* Roland T. Rust and Richard L. Oliver, eds. Newbury Park, Cal. Sage Publications, 1993.

must develop a keen understanding of how to improve service quality, and how to make sure that the business plan explicitly addresses customer needs and how they will be addressed.

WHY PRODUCT COMPANIES ARE SERVICES

It sounds strange to say that a product company is a service, but that is exactly what we argue. Let us consider, for example, a car manufacturer. One might be tempted to say that this company exists to sell cars. However, this is only part of the story. To see this, imagine what the company would be like if all it did was sell cars. Presumably, there would be an order clerk in a small office in Detroit, Tokyo, or Spring Hill, Tennessee, waiting for customers to arrive and order a car. There would be very few customers, because who would want to travel hundreds or thousands of miles to buy a car, unless the car were truly extraordinary? Thus some selling service is required. There should be a convenient dealership, with appropriate grounds and showroom (service environment). The demonstration and negotiation strategies should be well thought out (service design). For example, many car dealers are now adopting "haggle-free" pricing. This is a service design aspect. Also, the salesperson should be friendly and knowledgeable and treat customers with civility (service delivery). Thus you can see that this "product business" has all of the service components to worry about, and this does not even include the dealer's "service department," whose purpose is to repair and maintain the cars after they are sold.

The distinction between the product and service sectors can be quite arbitrary. Consider two companies, a car dealer and a car leasing company. The first is a product business while the second is a service business. Yet both companies exist to satisfy the transportation needs of customers, perhaps with identical cars, in almost the same way. In both cases, the customer will drive the car away and will send monthly payments to the seller (or lessor). The only difference is a piece of paper that says that in a few years the leasing company can have the car back.

Other businesses that are often (appropriately) thought of as services are strictly product businesses. For example, retailers sell physical products, as do restaurants. Even hotels often provide soap, shampoo, and other small items that become the property of the customer. Thus you can recognize that even hotels might be referred to strictly as product businesses. The product vs. service distinction has outlived its usefulness. You should instead recognize that all companies supply services, and that the components of these services must be improved and managed.

THE PHYSICAL PRODUCT

If business is viewed as a service, then all components of the service must be managed to meet customer needs. In the case of the physical product, product design must be customer-oriented. There is a well-developed method for doing this, called Quality Function Deployment (QFD). This approach became popular in Japan and was the source of much of its new product design success. It is a revolutionary approach, because it essentially says that marketing people and engineers should talk to one another.

Figure 1-7 shows the essence of the QFD approach, often referred to in the United States as the "house of quality."[32] Marketing's role in the QFD approach is to obtain the "voice of the customer." This involves identifying customer needs (e.g., styling, safety), prioritizing those needs, and examining existing competitive performance in meeting those needs. Meanwhile, engineering is identifying design attributes (e.g., air bags, reinforced construction) and is comparing the engineering performance of the firm's existing products with that of competitors.

The design process then combines the marketing side (voice of the customer) with the engineering side (voice of the engineer). First, the design attributes must be linked to customer needs. For example, if there are two air bags rather than one, how much does this affect the customer's perception of safety? Also, the design attributes may affect each other—often adversely. For example, if there are more air bags, the interior styling may suffer. Or if the doors have stronger reinforcement, this may make the car heavier and reduce gas mileage. These interactions

FIGURE 1-7   THE QUALITY FUNCTION DEPLOYMENT (QFD) APPROACH

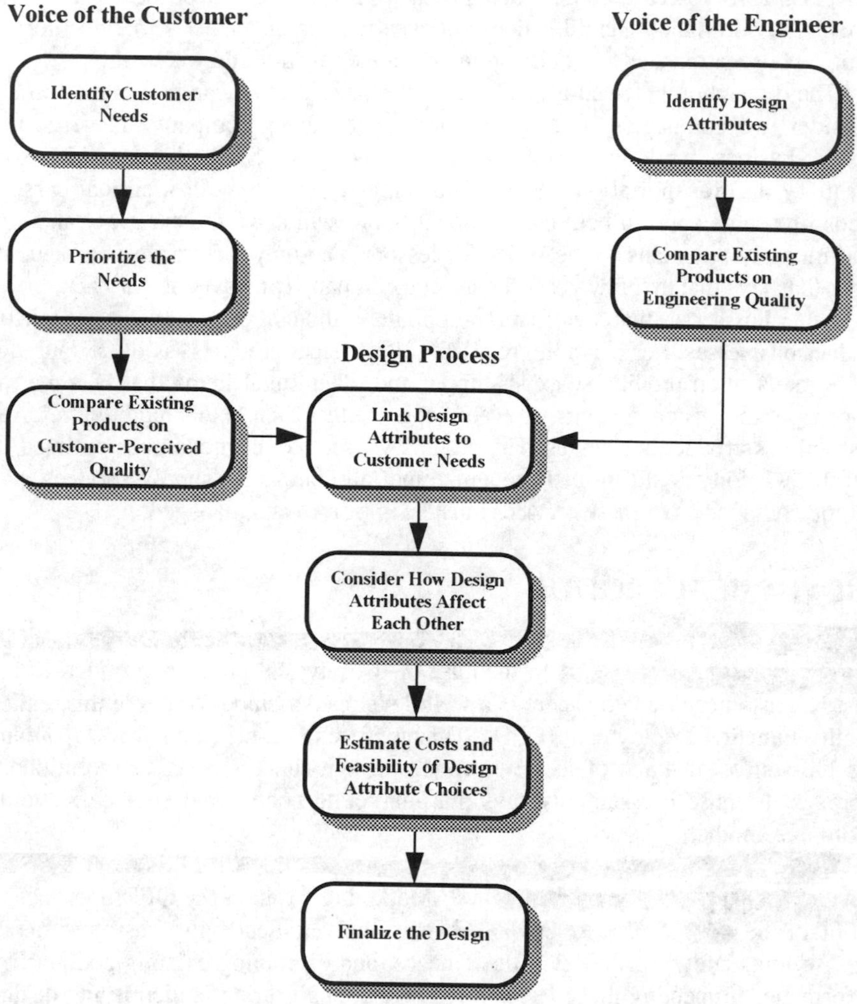

Voice of the Customer   **Voice of the Engineer**

```
┌──────────────────┐                              ┌──────────────────┐
│ Identify Customer│                              │  Identify Design │
│      Needs       │                              │    Attributes    │
└──────────────────┘                              └──────────────────┘
         │                                                  │
         ▼                                                  ▼
┌──────────────────┐                              ┌──────────────────┐
│  Prioritize the  │                              │ Compare Existing │
│      Needs       │                              │   Products on    │
│                  │                              │Engineering Quality│
└──────────────────┘                              └──────────────────┘
         │                    Design Process                │
         ▼                                                  │
┌──────────────────┐        ┌──────────────────┐           │
│ Compare Existing │        │   Link Design    │           │
│   Products on    │───────▶│  Attributes to   │◀──────────┘
│Customer-Perceived│        │  Customer Needs  │
│     Quality      │        └──────────────────┘
└──────────────────┘                 │
                                      ▼
                            ┌──────────────────┐
                            │Consider How Design│
                            │ Attributes Affect │
                            │    Each Other     │
                            └──────────────────┘
                                      │
                                      ▼
                            ┌──────────────────┐
                            │ Estimate Costs and│
                            │Feasibility of Design│
                            │ Attribute Choices │
                            └──────────────────┘
                                      │
                                      ▼
                            ┌──────────────────┐
                            │Finalize the Design│
                            └──────────────────┘
```

Source: John R. Hauser (1993), "How Puritan Bennett Used the House of Quality." *Sloan Management Review* 35 (Spring), pp. 61–70.

among the design attributes must be considered. Finally, after assessment of the costs and feasibility of the design attribute choices, the design is finalized.

Note that this approach explicitly considers customer needs. Design improvements that are "better engineering" but have little effect on customer needs are ill-advised. This error is what business writers refer to as "overengineering," and it is seen most often in technology-driven companies.

THE SERVICE PRODUCT

The service product includes everything that is *designed into* the service except the physical product itself. An example is the ordering procedure at a restaurant. There are many ways for customers to order. They may sit down at tables and order from a menu. They may stand in a cafeteria line and place the desired items on a tray. Or they may stand at a counter and order at a cash register. Notice that the service product says nothing about how well the design actually works. That is service delivery. You've heard the adage, "Plan your work, and work your plan." The service product is the result of "planning your work," and the service delivery is the result of "working your plan."

The QFD approach is useful here, just as it is for designing the physical product. For example, suppose you are planning a restaurant that will cater to lunch customers who are in a hurry. The design attributes, as in the previous section, might include the various ordering procedures. One of the key customer needs would be speed. It would soon be obvious that some ordering procedures (e.g., ordering from a menu) are slower, and thus would not be a good fit to customer needs in the target customer group. Once the needs of the customers are identified, it is often relatively easy to analyze how the design attributes will affect them. The key is to explicitly identify the customers' primary needs, and to explicitly consider how the design attributes affect them.

THE SERVICE ENVIRONMENT

The service environment is composed of the physical surroundings that facilitate the delivery of service. For example, the service environment of a ski area generally includes a mountain, a ski lift, and perhaps a lodge. The service environment is sometimes referred to as a "servicescape."[33]

Research indicates that besides being pleasant in its own right, the service environment can influence how customers respond to "critical incidents," service events that may result in the customer becoming dissatisfied (or sometimes delighted).[34] The service environment can also be used to signal the intended market segment and position the organization.[35] For example, a restaurant near a university campus might signal that it is catering to college students by putting college memorabilia and pictures of students on the walls. A car dealer might use its service environment to position itself as upscale by decorating its showroom in a tasteful and elegant manner.

There are three distinct elements that can be manipulated in the service environment: the ambient conditions, the spatial layout, and the signs and symbols.[36] The ambient conditions include things such as the lighting and background music. What may be appropriate for some businesses may be totally inappropriate for others, depending upon the market positioning. For example, bright lighting is appropriate for a fast-food restaurant but would be inappropriate for an expensive, romantic restaurant. The spatial layout can also have an impact. For example, Disney World found out that waiting lines seem shorter if the lines go around frequent turns, and there is some entertainment on the way. Signs and symbols are also important. For example, a car dealer selling American cars may fly a big flag to remind potential customers to "buy American."

THE SERVICE DELIVERY

Service delivery is what actually happens, not what was supposed to happen. That means that both good and bad surprises are possible. For example, consider a visit to a grocery store. The produce clerks may crush the lettuce (bad service delivery), or they may go out of their way to help you find the garlic cloves (good service delivery). The service design may be that the clerks are never supposed to crush the lettuce, and they are always supposed to be helpful. But what is designed does not always occur.

Most services are decentralized. Consequently, while the quality of the physical product can be relatively easily maintained at the factory through the careful use of statistical quality-control techniques, service delivery usually is much more geographically dispersed and much harder to monitor. Think, for example, of Wal-Mart, the huge retailer. Wal-Mart has stores all across the United States. If high-quality service delivery occurs, it is because of management at the local level. Of course, senior management can design the service environment and service product to obtain the best results, but ultimately the quality of service delivery depends upon the actions of (often low-paid) front-line employees at the local level.

SUMMARY

The justification for any quality-improvement program is its connection to the bottom line. However, experiences by such firms as Wallace Company and Florida Power & Light demonstrate that spending on quality is not a guarantee of profits. Therefore, firms need to determine their return on quality in order to best allocate their limited resources. The purpose of this book is to show managers how to measure ROQ.

Determining ROQ requires that companies understand the chain of events that leads from quality to profits.

Measuring the impact of quality on profits requires that companies determine the attributes customers use to evaluate their service and what factors drive customer satisfaction. The first step in this process is to view their organizations as services that are designed to meet customers' needs. This is reasonable, since all companies, whether or not a physical product is involved, have the service components of service product, service environment, and service delivery.

NOTES

1. Thomas Peters and Robert Waterman, Jr., *In Search of Excellence*. New York: Harper & Row, 1982.
2. P. Ranganath Nayak, Senior Vice President, Arthur D. Little, Inc., from a study of 500 executives conducted March 24, 1992.
3. Kevin Doyle (1992), "Who's Killing Total Quality?" *Incentive* (August), pp. 12-19.
4. Robert H. Schaffer (1992), "The Quality Quagmire." *CIO* (November 1), pp. 28-31.
5. Mark Ivey and John Carey (1991), "The Ecstasy and the Agony." *Business Week* (October 21), p. 40.
6. Mark Ivey and John Carey (1991).
7. L.M. Sixel (1992), "Quality-Award Winner Files for Chapter 11." *The Houston Chronicle* (January 30), Business Section, p. 1.
8. L.M. Sixel (1992), "Baldrige Winner Wallace Co. Sold to Louisiana Firm." *The Houston Chronicle* (August 5), Business Section, p. 2.
9. W. Earl Sasser, Jr., Christopher W. L. Hart, and James L. Heskett (1991), *The Service Management Course: Cases and Readings*. New York: The Free Press, pp. 427-444.
10. L.M. Sixel (1991), "The Quality Question: Have the Means Become an End for Many Firms?" *The Houston Chronicle* (October 20), Business Section, p. 1.
11. Jeremy Main (1991), "Quality Fever at Florida Power." *Fortune* (July 1), p. 65.
12. L.M. Sixel (1991), "The Quality Question: Have the Means Become an End for Many Firms?" *The Houston Chronicle* (October 20), Business Section, p. 1.
13. Robert Chapman Wood (1991), "A Hero Without a Company." *Forbes* (March 18), pp. 112-114.
14. Jeremy Main (1991).
15. "The Post-Deming Diet: Dismantling a Quality Bureaucracy." *Training* (February 1991), pp. 41-43.
16. William W. Arnold and Jeanne M. Plas (1993), *The Human Touch: Today's Most Unusual Program for Productivity and Profit*. New York: John Wiley & Sons, p. 11.
17. Joe Hall (1993), "Arnold's Style Cost Him His Job." *Nashville Business Journal* (May 24), Section 1, p. 3.
18. David A. Fox (1993), "Centennial Medical Center Names a New President." *The Tennessean* (May 30), p. E1.
19. Stephen B. Schwartz (1992), "Seven Milestones." *Executive Excellence* (May), pp. 15-16.
20. Bruce C. P. Rayner (1990), "Market-Driven Quality: IBM's Six Sigma Crusade." *Electronic Business* (October 15), pp. 68-74.
21. Karen Bemowski (1991), "Big Q at Big Blue." *Quality Progress* (May), pp. 17-21.

22. John F. Akers (1991), "World-Class Quality: Nothing Less Will Do." *Quality Progress* (October), pp. 26-27.
23. Suzanne Axland (1993), "NASA's Low Award Recognizes High Quality." *Quality Progress* (February), pp. 33-34.
24. Robert Buzzell and Brad Gale, *The PIMS Principles: Linking Strategy to Performance.* New York: Free Press, 1987.
25. L. Phillips, D. Chang, and R. Buzzell (1983), "Product Quality, Cost Position and Business Performance: A Test of Some Key Hypotheses." *Journal of Marketing* 47 (Spring), pp. 26-43.
26. E.B. Baatz (1992), "What Is Return on Quality and Why Should You Care?" *Electronic Business* 18 (October), pp. 60-66.
27. W. Edwards Deming, *Out of the Crisis.* Boston: MIT Press, 1986.
28. For a thorough review of this work, we refer the reader to Anthony J. Zahorik and Roland T. Rust (1992), "Modeling the Impact of Service Quality on Profitability: A Review," in *Advances in Service Marketing and Management,* T. Swartz, ed. Greenwich, Conn.: JAI Press, pp. 247-276.
29. See Arch Woodside, L. Frey, and R. Dely (1989), "Linking Service Quality, Customer Satisfaction and Behavioral Intention." *Journal of Health Care Marketing* 9 (December), pp. 5-17; William Boulding, Ajay Kalra, Richard Staelin, and Valerie A Zeithaml (1993), "A Dynamic Process Model of Service Quality." *Journal of Marketing Research* 30 (February), pp. 7-27; Roland T. Rust and Anthony J. Zahorik (1993), "Customer Satisfaction, Customer Retention and Market Share." *Journal of Retailing* 69 (Summer), forthcoming; and Eugene C. Nelson, Roland T. Rust, Anthony J. Zahorik, Robin L. Rose, Paul Bataldan, and Beth Ann Siemanski (1992), "The Effect of Patient Satisfaction on Hospital Profitability." *Journal of Health Care Marketing* 12 (December), pp. 6-13.
30. Roland T. Rust and Richard L. Oliver, "Service Quality: Insights and Managerial Implications from the Frontier," in *Service Quality: New Directions in Theory and Practice,* Roland T. Rust and Richard L. Oliver, eds., Newbury Park, Calif.: Sage Publications, forthcoming.
31. John E.G. Bateson (1989), *Managing Services Marketing,* 2d ed., The Dryden Press, and Steven M. Shugan, "Explanations for the Growth of Services," in *Service Quality: New Directions in Theory and Practice,* Roland T. Rust and Richard L. Oliver, eds. Newbury Park, Calif.: Sage Publications, forthcoming.
32. See John R. Hauser and Don P. Clausing (1988), "The House of Quality." *Harvard Business Review* 66 (May-June), pp. 63-73; John R. Hauser (1993), "How Puritan-Bennett Used the House of Quality." *Sloan Management Review* 35 (Spring), pp. 61-70; and Abbie Griffin and John R. Hauser (1992), "Patterns of Communication Among Marketing, Engineering and Manufacturing—A Comparison Between Two New Product Teams." *Management Science* 38 (March), pp. 360-373.
33. Mary Jo Bitner (1992), "Servicescapes: The Impact of Physical Surroundings on Customers and Employees." *Journal of Marketing* 56 (April), pp. 57-71.
34. Mary Jo Bitner (1990), "Evaluating Service Encounters: The Effects of Physical Surroundings and Employee Responses." *Journal of Marketing* 54 (April), pp. 69-82.
35. Bitner (1992).
36. Bitner (1992).

METHODS OF LISTENING TO THE CUSTOMER

THE IMPORTANCE OF LISTENING TO THE CUSTOMER

What does "quality" in a service mean? Engineers who design and manufacture tangible products often use a "conformance to specifications" definition of quality—in other words, a measure of whether the product performs (or measures or weighs) exactly as it was designed. For example, Motorola, Inc., a recognized world leader in manufacturing quality, has instituted a program called "Six Sigma Quality" in its manufacturing operations, which means that its systems must be under such control that fewer than 3.4 operations in a million will fall outside specified tolerances.

Some service firms have also adopted this goal. For example, American Express Company's Travelers Cheque Group has adopted a six-sigma standard for many of its customer service measures. However, this kind of precision is not really possible for most services to achieve, particularly for those in which the customer plays a key role in the completion of the service. Moreover, services are performances, with many intangible, difficult-to-measure components. A customer of a tax-preparation service may agree that her tax form was filled out without mistakes—the firm's quality objective—but be dissatisfied because she found the tax preparer's personality abrasive. This customer would not rate the firm as providing excellent service, even though it is meeting its goals for precision and accuracy.

In fact, defining quality as conformance to specifications leaves out a key decision-maker, the customer. It is the customer's specifications, not those of the engineering department, that must be met for the service to be deemed successful and worthy of repeat purchases or recommendations to friends. Therefore, although the term "quality" has many different meanings, there is a general convergence toward a definition that reflects the extent to which a product or service meets or exceeds customer expectations for it. The importance of this definition is its assumption that quality is ultimately a subjective determination by the customer. Precision production processes may be necessary to deliver consistently high quality, but if customers don't notice, or if they are really concerned about other aspects of the service, then high precision doesn't guarantee that customers will consider the quality good. A firm must be careful that the internal measurements it takes to monitor quality (e.g., average waiting time or number of keystroke errors or number of rings before a call is answered) are related to the measures of perceived quality that customers use.

Therefore, it is essential to have a thorough understanding of customers' views of a service before one is able to measure its quality, to adequately fix problems, or to improve it significantly. And customers may have very different views of the service from that of the provider. For example, until recently, when competition began forcing the medical profession to be more customer-oriented, doctors were notorious for overlooking the customer's perspective on the quality of care provided. (And not all doctors have improved!) A doctor often felt that his education from the most prestigious medical schools and the expert and professional manner he used while interacting with a patient guaranteed quality health care. On the other hand, the patient had to deal with a rude receptionist to get an appointment at an inconvenient time, difficulty in parking, out-of-date magazines to read during long waits in the waiting room and in the examining room to be with the doctor for five minutes, and no help in filing insurance claims afterward. To the patient, the doctor may have seemed competent but cold, and the entire experience was rated as trying and wasteful. Two very different views, and a great marketing opportunity for a doctor with sensitivity to customer concerns!

Part of the explanation for such very different evaluation standards is that customers often have a great deal of difficulty judging service quality. The quality of such services as medical operations, auto repair, or legal representation may be very hard for the average person to judge. In such cases, customers tend to rely on whatever cues they can understand—cues that may have little or nothing to do with the core service being provided. For example, hospital administrators generally find that patients will give high ratings to the overall quality of their care during even the most difficult medical procedures if the staff has smiled at them frequently during their stay. It's the health-care equivalent of the automakers' concern that the car doors sound solid when slammed, since customers tend to judge the overall quality of their incomprehensibly technical cars by such easy-to-read clues.

Customers also tend to use different language from that of the service providers. Successful service firms need to know this vocabulary, because it often carries important information about customers' subjective feelings and can also be used in communicating with the customer. For example, a concern that retail banking customers frequently express when judging the service of their bank is whether the bank "listens to my needs." The implications of this term for the design of new services or the improvement of old ones isn't obvious to the outsider. It certainly doesn't relate immediately to such concerns of the bank operations officer as turnaround time, average queue length, or statement accuracy. But over time, bankers have come to understand what customers mean by that vague expression and have developed services and training programs for staff personnel designed to improve customer perceptions of the bank's performance on this important aspect of the banking relationship.

The moral of this section is that service quality (and product quality as well) is in the mind of the customer, which means that measuring quality requires talking with customers. In this chapter, you will learn some methods of exploratory research that have been developed to explore the customer's view of a service.

THE STRUCTURE OF A SERVICE

Think of a service you experienced recently that you found dissatisfying. Chances are that not everything about it was terrible, even though your overall opinion of it was low. You will probably admit that some aspects were done well, just not well enough—or not the right ones, or not enough of them. The company might have been able to change your mind about the overall service not by fixing every problem but by doing better in one or two key areas. Your overall opinion of an experience doesn't usually require that everything about it be satisfactory. Rather, overall satisfaction or dissatisfaction is the result of weighing positive and negative experiences and coming to some net evaluation. Before you can determine how to increase customers' satisfaction scores profitably, you need to know what these major components are, how relatively important each is to the overall impression, and, perhaps, what subcomponents determine satisfaction with each of the components.

Figure 2-1 shows the standard conception of how customers organize information about a service. The overall evaluation is determined as a weighted average of the evaluation of the service's various processes or components. For example, patient evaluations of hospital visits may depend upon their impressions of various components, such as admission, testing, the nursing staff, the doctors, the accommodations, rehabilitation services, the discharge process, and the billing operation. Each of these components can be broken down further to help diagnose specific problems or opportunities for superior service. For example, patient satisfaction with billing could depend to varying degrees on the accuracy of the bill, the willingness of the hospital to provide clear explanations of items, the readability of the statements, etc.

FIGURE 2-1   FACTORS DRIVING CUSTOMER RETENTION

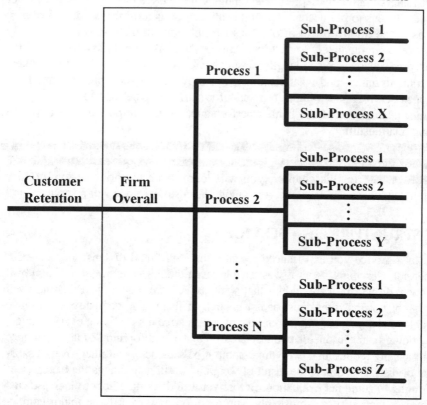

Customer Satisfation With the Various Elements

How does a firm develop the customers' list of components and subcomponents of a service? In this chapter we will describe several exploratory research techniques whose purpose is to define the service dimensions that determine customer satisfaction. The goal in using these techniques is the generation of a list of 200 to 400 customer needs, which can be grouped together to form a hierarchical structure of the services components, such as that in Figure 2-1. Methods for performing that grouping will be described later in the chapter. Don't be concerned at this time that all the needs on the list are truly important determinants of customer satisfaction. Formal statistical tests on the data will determine that later. For now, your goal is simply to gather all the potential needs you can, so that key determinants of service quality don't get overlooked.

Customer needs can be grouped into three different categories, each of which has different strategic implications.

1. **Basic attributes** are those assumed to be present in any service. For that reason they may not always show up in customer discussions, even though they are essential. A firm that doesn't offer these attributes could not remain in business. Examples include safe operations by airlines, working restrooms in a restaurant, and clear connections with a long-distance phone company.

2. **Articulated attributes** are those which customers generally mention as desirable or determinant in their choice of a service. Poor performance on these attributes will lead to customer dissatisfaction, and good performance will lead to satisfaction. They are the bread-and-butter attributes on which firms compete and differentiate themselves: the bank with the friendliest tellers, the auto repair shop that fixes your muffler on time and at the price promised.

3. **Exciting attributes** are those which would delight and surprise customers if they were present. Delivering these attributes is the key to building strong loyalty among customers, but they are also harder to identify. Customers don't expect them, so it takes careful questioning to get customers to suggest them. Their power is also fleeting. Attributes that are stunning today become commonplace tomorrow. Auto Zone, a retailer of new replacement parts for automobiles, set new standards in the areas of inventory levels and free advice for customers. In time, however, that positioning came under attack from Pep Boys, which carries larger inventories and also offers service bays and installation. Auto Zone must find new ways to differentiate itself.

USING A GROUP OF SUBJECTS: FOCUS GROUPS

Perhaps the most widely used technique for listening to the voice of the customer is the focus group. A focus group is a moderated discussion of a predetermined topic by 6 to 12 participants. The idea is to generate a free-flowing discussion about the service among members of a target group, using interaction among participants to relax them and to trigger ideas. The discussion is typically recorded on both video- and audiotape for later transcription and analysis. Written transcripts are essential when focus group discussions are used to generate lists of customer needs, because they must be examined in detail by several analysts to identify as many ideas as possible. Many focus groups are also conducted behind soundproof, one-way glass so that observers from the company can watch and discuss the proceedings as they happen.

The success of a focus group depends on the skills and preparation of the moderator, who must put the participants at ease and keep the discussion focused on the topic at hand without suppressing people's willingness to speak or injecting his or her own opinions. This can be a difficult balancing act, as one must be able to keep the occasional overly talkative participant from dominating the discussion while encouraging all participants to speak their minds. It is also the moderator's job to see that only one person speaks at a time, or else valuable comments can be misunderstood or missed. This requires tact as well, because as enthusiasm builds among participants, there is often a tendency for side conversations to occur. They must be discouraged or, better, channeled into the main conversation, without dampening the group's level of excitement.

The moderator should have a loose script prepared in advance listing the key topics to be covered during the session. As the group runs out of fresh things to say on one topic, the moderator should be prepared to introduce new topics, or to ask follow-up questions to encourage discussion and to stimulate new ideas among all group members.

Practical Considerations

Focus group members are usually recruited by telephone. They are offered refreshments and cash, with the amount depending upon the target market sought, for a session of 90 minutes to two hours. For many recruits, the chance to discuss their experiences with a particular product or service is seen as an interesting experience. In spite of these inducements, some individuals inevitably fail to show up at the appointed time, so experienced focus group recruiters typically recruit 10 to 12 people in order to ensure 8. Should more people arrive than are needed, the extra people must still be paid for their trouble, but can be sent home.

Professionally designed focus group facilities, with microphones, video cameras, multiple recording facilities, one-way glass, reception area, etc., are obviously the most desirable setting but can be rather expensive to rent. Alternatively, a room with a large table will do, provided that participants can be put at ease in the surroundings and clear recordings of the session can be made.

Limitations

It cannot be stressed too much that focus groups are only exploratory, and that their findings cannot be statistically projected to the population at large. Findings from focus group sessions can't be relied upon until they have been formally tested using statistically analyzable surveys. This fact tends to become obscured as managers become engrossed in viewing the sessions, reading the transcripts, and studying the reports. There is a tendency to make too much of comments made in focus groups and to treat them as representative of the general public. This is dangerous for several reasons.

First, focus groups rarely constitute a random sample of the population of interest. Recruits are usually just a "convenience" sample of the target group, people who were easy to reach, had some time to spare, and were able to get to the sessions. Therefore, the sample cannot be considered representative of the population at large. But a focus group's purpose goes far beyond trying to identify how the population feels about a specific, predetermined issue. It seeks to sample the set of all ideas in the minds of the population—a much larger task. However, the unstructured nature of focus group discussions and the vagaries of group dynamics mean that there is little chance that the concerns and ideas expressed represent a random sampling of those of the population at large.

For example, in a focus group on checking accounts, someone might mention the designs available on checks at her bank, perhaps due to some experience that day. Other customers may pick up on the topic simply because they have something to say about it, and 15 minutes or more of a 90-minute session will be devoted to a topic that actually has very little bearing on anyone's satisfaction with a checking account. Without strong justification for intervening, however, the moderator must let such a discussion run its course, on the chance that some vitally important but previously overlooked attribute of checking accounts will be revealed. Subsequent analysis, perhaps including a question on the importance of check patterns on a formal survey of bank customers, will help to determine whether the issue was due to momentary group sympathies or is a real issue.

One or two participants with dominant personalities can also sway a discussion into unrepresentative areas through social pressure. For this reason, focus groups are of limited use in getting candid reactions to socially risky concepts or in eliciting aesthetic judgments. If a dominant member of the group starts the discussion by declaring an introduced concept to be in poor taste, or something only a fool would buy, it tends to severely inhibit those who were about to declare their approval.

Finally, using the focus group format to identify service attributes relies on strength of numbers, rather than on the depth of questioning, to produce a full list. In a two-hour session with 10 participants, each participant has an average of only 12 minutes to discuss all he or she feels about the service. This limited "air time" provides very little chance to probe deeply into any one customer's views. As a result, the attributes revealed by any particular individual may be the most obvious ones, or those "closer to the surface."

A variant on the focus group is the mini-group, consisting of just three or four participants. This format provides some of the benefits of interaction from group dynamics but allows more time to dig in deeply. However, the smaller group also makes the experience less exciting for participants, and may not stimulate as much discussion. If true in-depth follow-up is desired, the best format is probably the one-on-one interview.

ONE-ON-ONE METHODS: THE IN-DEPTH INTERVIEW

Subjects for one-on-one interviews are recruited and rewarded similarly to focus group participants. These sessions must be recorded for future transcription but are generally not observed in progress, so smaller, more informal facilities can be used. Every effort should be made to make the subject comfortable and at ease, because this format does not offer the strength in numbers or moral support for one's opinions that the larger focus group can provide.

Interviewing 10 people one at a time obviously takes longer than running a focus group and also puts greater demands on the interviewer. Without the stimulation of group experiences to trigger ideas or the ability of subjects to sit back and think while others speak, the moderator must use a more structured approach to help subjects explore their experiences.

A one-on-one session usually takes about one hour, and in addition to an unstructured discussion of the topic at hand, researchers sometimes use one or more exercises designed to stimulate thinking about service attributes. Some of the most popular are described below. Each of these exercises can be used to stimulate focus group discussions as well.

EXERCISES TO ELICIT SERVICE ATTRIBUTES
Triadic Sorting (Kelly's Repertory Grid)

This well-established marketing research technique is specifically intended to force customers to think about and express the product/service attributes that they use when choosing among competitors in a category.[1] The five steps in the exercise are:

1. A list of competitive alternatives in the market is prepared and printed on numbered cards.
2. The subject is asked to go through the cards and remove any alternatives with which he or she is completely unfamiliar.
3. Three of the cards are then presented to the subject according to a prespecified sequence. The subject is asked to think of any basis on which two of the three are similar to each other but different from the third.
4. The exercise is repeated for each group of three in the prespecified sequence. In each case the subject is asked to think of a new way in which two are similar to each other and different from the third.
5. The lists of attributes used to group and separate the competitive alternatives is recorded until the subject is unable to think of new ones.

The list of attributes generated is unlikely to be exhaustive, but the exercise can generate some surprising insights and should stimulate thinking and discussion of attributes.

Customer Service Scripts

Physical products are often treated by marketers as bundles of attributes which are all experienced together. For example, a car has a color, a certain number of doors, horsepower rating, trunk capacity, etc. This model works because it adequately describes the way most customers view goods—as things, single entities to be used or consumed as a whole. But services are generally different from goods in that there is a temporal sequence to the component parts. Services are performances that often occupy time as well as space. They are experienced as well as consumed. Therefore, a list of a service's attributes, no matter how complete, cannot fully describe the customer's experience with it. The service is a sequence of events whose order and individual qualities define the service.

For services in which customers play an interactive role, such as ordering materials from a supplier or dining in a restaurant, the customer typically has a mental script that allows him to act appropriately throughout the performance of the service while evaluating the service experience. He has expectations about the order and duration of each phase of the service, and he knows what to say and do in response to each action by the service provider. Deviations from the expected script can lead to discomfort and dissatisfaction.

For example, the restaurant patron will not evaluate the experience solely by whether the food was good and the staff polite, but by how comfortable the entire experience felt. Partly for this reason, restaurants tend to follow very rigid scripts. From the opening question at one's first arrival at the restaurant ("How many in your party?" or "Do you have a reservation?"), the interchange between patron and staff seems unvarying across a wide range of restaurant types and geographic regions. Experienced restaurant-goers "know their lines" when they go out to eat, and can become confused or think poorly of the restaurant if the flow of events and the interchange with the staff is not as expected. For example, one is pleased to have the waiters hovering nearby during certain stages of the meal. At other times, their presence could be considered intrusive.

Therefore, it is critical that service firms identify the scripts that their customers bring to the service. The "scenes" of the script usually correspond to the processes of the service. The process of eliciting scripts from customers merely involves taking them through a service experience and probing in detail for what they expect, or feel is expected of them, at each phase of the experience. To keep the discussion realistic and concrete, the interviewer should concentrate on a specific occasion—perhaps the last time the subject used the service.

The Critical Incident Technique[2]

The critical incident technique (CIT) is a one-on-one exercise for customers to elicit details about services that particularly dissatisfy or delight them. Data are generally collected from large enough samples of subjects that patterns of responses can be identified. In particular, the interviewer asks probing questions:

- What makes a service encounter particularly satisfying to customers? Do specific events constitute a satisfying service experience? What do contact employees do that cause these events to be remembered favorably?
- What makes a service encounter dissatisfying to customers? What events occur, and what do contact people do that causes them to be remembered unfavorably?

■ Are the components that determine satisfactory and unsatisfactory encounters related—e.g., opposites or mirror images of each other?

A typical interview script asks a subject to recall an incident when, as a customer of the industry (firm) in question, he had a particularly satisfying (dissatisfying) interaction with an employee. The subject is asked to fully describe the situation: When did it happen? What specific circumstances led up to the incident? What mood or frame of mind were you in? Exactly what did the employee say or do? What resulted that made you feel the interaction was satisfying (dissatisfying)?

Transcripts of the interviews are then compared to identify common problems or sources of delight. Specific points mentioned by the sample of customers provide a list of the key processes and dimensions of the service as customers see them. The specific sources of problems and delight can also be grouped by process and dimension to highlight those components of the service that need attention or where greater customer loyalty can be achieved.

Laddering[3]

Laddering goes beyond simply identifying the attributes customers use to classify products. It seeks a deeper understanding of how product attributes are associated with consumers' personal beliefs and goals. As such, it provides more than just a list of attributes; it gives insights into why the customer thinks they are important. It is based on a psychological theory called Means-End Theory,[4] which deals with the connections between product attributes (the means) and the customer's personal values that the attributes reinforce (the end).

The interviewer uses a structured series of probing questions, typified by the question "Why is that important to you?" The procedure is as follows:

1. An important attribute of a service, perhaps one identified by the Kelly grid procedure, is selected. The subject is asked which extreme pole of the attribute he or she prefers. For example, in a study of automobiles, engine horsepower may be selected. The subject would then be asked if he or she preferred automobiles with more or less horsepower.
2. The subject is asked why he or she chose that pole. This can lead to distinctions among services based on the different reasons customers purchase them and the different consequences they produce.
3. The same procedure of questioning can be continued until customers ultimately distinguish among services not because of their obvious attributes, but based on the personal values they reflect and the consequences that result from using them.
4. When the subject can no longer provide answers, the exercise ends.
5. If the service can be used for different usage occasions, each use can be pursued separately.

Here is a typical ladder obtained from a secretary discussing why she would prefer to use an overnight package delivery service that has drop boxes available:[5]

The laddering technique can provide deeper insight into the positioning of a service in the minds of its customers. This depth insight is useful not just for writing compelling advertising copy. Knowing why customers care about certain attributes may suggest the kinds of quality improvements that will be most meaningful to customers.

LEADING-EDGE USER STUDIES[6]

Questioning customers about their experiences with your service tends to provide mostly articulated attributes, those features that describe current market offerings. It is harder to get customers to imagine and to articulate exciting service features, those that would exceed their expectations. And yet, rather than rely entirely upon the creativity of your R&D effort to provide new products for the future, you can look to your own customers for guidance about what the market wants.

The key is to identify an important market or technical trend, based on a careful analysis of the business environment: look for economic, demographic, technological, legal, and political changes that can affect your customers and the way they use your services.

The firm must next identify those users of the service who lead the trend in terms of their experience with the service and the intensity of their need to adopt it. So-called leading-edge users have two distinguishing characteristics:

1. Their needs for service enhancements are similar to the marketplace as a whole, but they arise months or years before the rest of the marketplace faces them; and

2. They will benefit significantly by finding solutions to those needs.

The leading-edge users must be carefully studied to find out the problems they have with the current market offerings, what solutions they suggest, and what "wish lists" they have for service suppliers. Very often these intense users are knowledgeable enough to have developed some of their own solutions to their problems, and these home-grown solutions can offer excellent guidance in developing features that should be incorporated into future offerings.

Finally, the suggestions obtained from the lead-user data must be projected onto the market as a whole to determine the extent to which they really anticipate the coming general demand or are specific to these heavy users.

DIRECT OBSERVATION

The concept of Management By Walking Around (MBWA) has been practiced by good companies for many years, but it gained its capitalized name and a good deal of publicity in Peters' and Waterman's book *In Search of Excellence*.[7] MBWA refers to various methods of direct informal observation intended to supplement and breathe life into the statistics that most managers use to monitor the progress of their businesses. The basic idea is that managers must get out of their offices and meet their customers, with the goal of experiencing the business through their customers' eyes. MBWA differs from most of the other techniques described in this chapter in that, rather than professional market researchers conducting interviews and then distilling them into reports for management consumption, the contact is by managers and line workers themselves.

For example, some firms, such as Apple Computer, require (or at least encourage) company executives to answer telephone calls from complaining customers.[8] At other companies, executives are expected to serve periodically on the front lines of the firm's service outlets. Some companies regularly visit their customers to discuss how their relationships are holding up, or they invite customers for detailed visits of the firm's facilities and discussions about what it is like to do business with the firm.

This hands-on understanding of customer needs and desires is important for more than just top executives. Quality is ultimately in the hands of the hourly workers who perform the service, and so they, too, need to experience the service from the customers' viewpoint. For firms that sell to other businesses, Peters and Austin[9] recommend sending hourly workers to meet their counterparts at customer firms to hear how their work is being received. They even cite firms in which managers perform the tasks of hourly workers at a customer firm's company to gain a better understanding of their service needs. Customer letters of both praise and complaint should also be made widely available for all employees to see.

Another useful method for appreciating the customer's point of view is the exit interview—i.e., interviews of dissatisfied customers who have recently terminated their relationships. Although the discussion can be painful, it is very important to know what aspects of the service cause customers to defect. The primary purpose of these interviews is to gather information, and they should be treated as such. Their value as opportunities to salvage the relationship is often considerably lower.

MBWA activities are often cited as valuable tools to stimulate a firm's employees to take quality more seriously. Also, the insights they provide to managers will give them a much better understanding of how customers perceive the structure of the service.

Some Practical Issues

The goal of all these exercises is to compile a list of the drivers of satisfaction that is as complete as possible. If critical items are omitted from the list, you may overlook important sources of problems or cost-effective solutions for quality enhancements. But how effective are the above research methods at generating all of the major attributes of a service? A partial answer to this question has been provided by Griffin and Hauser,[10] who have performed a number of studies on the effectiveness of these techniques.

As one would expect, the more focus groups or in depth interviews that are conducted, the greater the number of attributes one is likely to generate. But these sessions take time and can be extremely expensive when one factors in the costs of management time spent on viewing interviews, reading transcripts, and discussing the data. Therefore, a tradeoff becomes necessary between the mounting costs and the decreasing amount of additional information obtained from doing more interviews. Griffin and Hauser directly addressed this tradeoff in a series of studies of exploratory methods applied to two product categories, a proprietary computer product and consumer food-carrying devices.

In a comparison of the ability of focus groups vs. one-on-one interviews to generate lists of customer needs, the authors discovered a surprising relationship. As expected, a given number of two-hour focus groups generated, on average, more needs than the same number of one-hour one-on-one interviews. The ratio of needs generated appeared to be constant: consistently, twice as many one-on-one interviews as focus groups were required to produce the same number of needs. However, since each focus group lasted twice as long as each in-depth interview, the two types of sessions each required the same total amount of time. The results suggest that it was total interview time, not the number of participants, that determined the number of needs generated. Maybe this shouldn't be too surprising in light of our earlier caution about the limited ability of focus groups to probe as deeply into any one individual's perceptions. However, this was a very limited study, and it would be foolhardy to claim the discovery of a new natural law on this slim evidence. The results do give researchers some guidance in designing research plans, but they should be questioned and tested until more data come in.

Another useful phase of the Griffin and Hauser study was an investigation into the average number of interviews necessary to generate a complete list of needs. They found that as the number of interviews increased, the total number of needs revealed seemed to approach a finite upper limit. Then, by fitting a theoretically reasonable curve to the increasing number of needs produced by various numbers of interviews, it was possible to estimate that upper limit, which represented the probable total number of needs. Using this value as the total number of needs, it was then possible to measure the average percentage of total needs generated by various numbers of interviews. For example, they found that it took five interviews, on average, to generate 50 percent of total needs and the equivalent of 25 interviews to get 98 percent coverage. The multiple sessions did not produce a great deal of duplication of needs among interviews. In fact, the bulk of the needs generated were each mentioned in fewer than 20 percent of the interviews. Getting a complete list of attributes requires more than just a few sessions.

A third insight in the Griffin and Hauser studies is that a single reader is unlikely to find all of the needs contained in the transcripts of focus groups and one-on-one interviews. In fact, trained analysts were each able to identify an average of only 54 percent of the needs contained in a set of transcripts, with a maximum of 68 percent. Analysts' preconceptions of what needs are and the inability of some analysts to distinguish between the nuances of certain statements may be partly responsible. The studies found that an average of seven analysts was necessary to identify 99 percent of the needs contained in the transcripts.

An important lesson from this research is that complete identification of the customer needs that define a service is not easy or cheap. The data used in these studies represented two rather narrow categories and have limited application to others, but the results show that it is possible to measure the cost tradeoffs one makes in generating data on customer needs. It is comforting to know that the studies also found that recent improvements in interviewing techniques are increasing the efficiency of this activity.

SECONDARY SOURCES OF SERVICE ATTRIBUTES

The attribute-generating exercises described in this chapter are only exploratory, and there is no guarantee that they will cover all the important aspects of a service. Therefore, before a formal customer questionnaire is finalized, the analyst is advised to look to other sources for possible missing attributes.

The most obvious source is management's common sense about the service developed through experience, including the heightened customer sensitivity developed through Management By Walking Around. The list should be reviewed by management to see if any key areas have been overlooked, particularly those where problems arise and diagnosis and monitoring are desired.

Trade publications occasionally have articles giving advice on quality in their particular industries, including lists of areas to be monitored. Since customer satisfaction questionnaires are often mailed out, it is often quite easy to obtain those used by competitors and by firms in other industries with similar services. These questionnaires can be checked for possibly overlooked service dimensions.

One final source of service attributes worth mentioning is the SERVQUAL model of Parasuraman, Zeithaml and Berry (PZB).[11] These authors conducted focus groups and then formal surveys of customers in several different service industries to develop lists of attributes that define service quality in general. The lists were condensed by correlational analysis into five major categories. PZB describe them as follows:

Tangibles	The appearance of physical facilities, equipment, personnel, and communications materials
Reliability	The ability to perform the promised service dependably and accurately
Responsiveness	The willingness to help customers and to provide prompt service
Assurance	Knowledge and courtesy of employees and their ability to convey trust and confidence
Empathy	Caring, individualized attention the firm provides its customers

Customer assessment of each of these dimensions is measured by comparing scales of customers' expectations and actual experiences on a battery of 22 items, approximately four items per dimension. PZB claim that Reliability appears to be the most important service dimension for customers across many industries in which they have applied the SERVQUAL methodology.

The existence of these dimensions is somewhat controversial among some researchers. Some have criticized the statistical methodology used to identify them.[12] It is also important to remember that the list is intended to describe the dimensions of quality that are common to all services, and is therefore unlikely to encompass all the special properties of any particular service industry. The SERVQUAL dimensions are also like product attributes—static descriptors of the service rather than components of a dynamic performance. Nevertheless, the five areas have been well accepted by service industry managers as having strong face validity, and no list of customer needs should be considered complete until it has been checked for representation of the SERVQUAL dimensions. For example, one should check the dimensions that describe each process in Figure 2-1 to see if any of the SERVQUAL dimensions, such as Tangibles, need to be represented in some way on the list.

ORGANIZING THE LIST OF ATTRIBUTES

Expert Analysis

Once the list of several hundred customer needs has been generated, it must be organized in ways that reflect the customers' perception of the structure of the service, and that give management a diagnostic tool for measuring and improving key components of the service.

One method, considered to be one of the basic components of Japanese management, is the development of a hierarchical tree diagram of the data using "K-J analysis." In this exercise, a multidisciplinary team of experts organizes the list of needs by group consensus. It uses a bottom-up approach, organizing the most detailed levels of needs and then seeking higher levels of organization in those lower-level groupings. This method is supposed to reduce the influence of the experts' preconceptions about what the structure should look like.

The exercise begins with the complete list of expressed needs, often reworded to express only positive sentiments. Team members are asked to cluster these data into groups of similar topics and to assign a heading that broadly describes each group. The individual statements represent the lowest level of details, which correspond to design details of the service and form the outermost branches of the tree diagram. At the next stage, these headings are themselves grouped into similar categories—the tactical level of details. These categories are once more clustered into strategic groupings, if possible. Finally, the structure should be assessed for completeness. If management feels that the structure is missing what it knows to be important levels of detail or other branches, it is appropriate to add them to the tree diagram at this stage.

The resulting structure might reflect the service's attribute structure, or perhaps the temporal sequence of processes that make up the service. The key is to find in the data the structure that most closely describes the customers' view.

Using Customers to Do the Groupings

In spite of the hope that K-J analysis will be free from the experts' preconceptions, the studies of Griffin and Hauser, cited earlier, also found a production bias in the structures developed by experts compared to those of customers. Customer-based groupings of needs tended to reflect products' functions and uses, while experts tended to group products more by technology. The experts themselves conceded that the customer-based structure had stronger face validity as a description of customer behavior than their own.

Griffin and Hauser's method for having customers organize the list of needs combined the individual judgments of a sample of subjects rather than a group consensus. It also allowed them to have each subject sort far fewer than the full set of several hundred needs—an absolute necessity to getting subject cooperation. The steps used were as follows:

1. Each customer was given a deck of cards, each bearing one item from the list of customer needs. (By using multiple subjects, and by carefully choosing overlapping subsets of needs determined according to an experimental design, it is possible to keep the individual decks small. This requires computer-based clustering techniques.)
2. Each subject was asked to sort the cards into piles that represented similar needs, but so that the piles were somehow different. The number of piles to use was not specified.
3. From each pile, the subject was asked to choose one card that best represented the needs in the pile. This need was called an "exemplar" for the pile.
4. The analyst then compiled a co-occurrence matrix, a square matrix with as many rows and columns as there were needs (see Figure 2-2). The matrix was used to tabulate the number of times subjects grouped a particular need with the other needs. (The analysis must compare these numbers for different combinations of needs, so in designing the decks to give to different subjects, it is essential that each pair of needs occurs in an equal number of decks.) Note from Figure 2-2 that because the same headings are used for both columns and rows, the cells in which row and column contain the same headings (the diagonal line in the matrix) will always equal the number of subjects used because, for example, Need 1 will always be grouped with Need 1. Also, each half of the matrix, as separated by this diagonal line, will be a mirror image of the other. Therefore, only half of the matrix needs to be used when tabulating the results.
5. The co-occurrence matrix was now interpreted as a similarity matrix. The higher the number in each cell, the more often needs were grouped together by subjects, and the stronger the argument for grouping them together near the ends of the tree branches. The mathematical techniques that can perform such similarity-based grouping are called cluster analysis. By applying cluster analysis to the matrix, a hierarchical tree structure of needs was developed that reflected the joint decisions of the subjects. Those needs that were chosen most often as exemplars could be used to name the clusters.

While the customer-based method produced superior results in the experimental setting, the requirement of computer-based clustering routines makes it far less accessible than K-J analysis, which requires no computer and offers an opportunity for managers to add in detail with customer statements of their perceptions of the service—a benefit in its own right. If the experts performing the K-J analysis have been practicing Management By Walking Around, as recommended earlier, they might be better attuned to customer thinking than those in the Griffin and Hauser

FIGURE 2-2   A CO-OCCURRENCE MATRIX FOR GROUPING CUSTOMER NEEDS

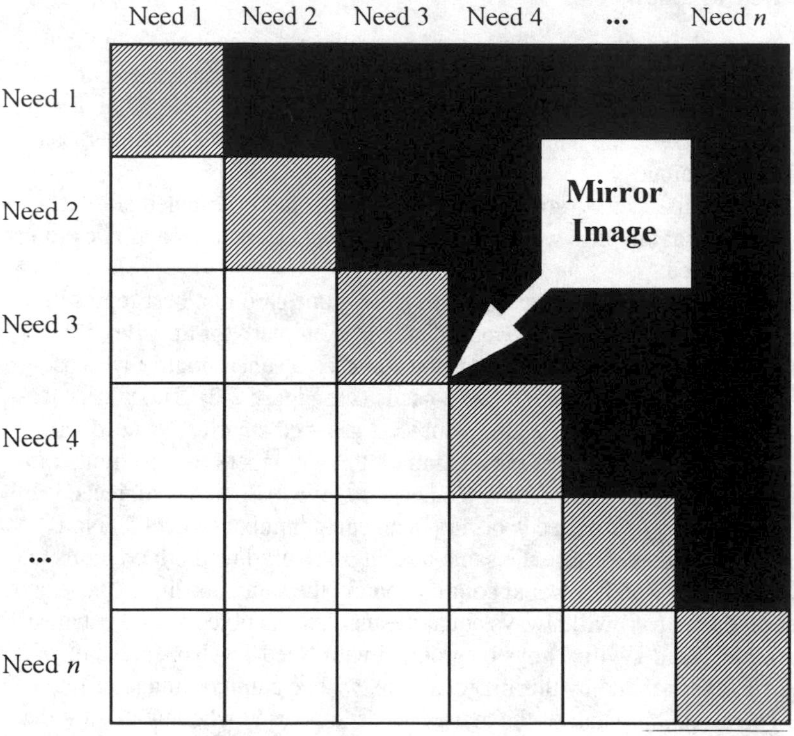

studies were, and may therefore produce hierarchies that more closely reflect customer thinking.

SUMMARY

Quality reflects the extent to which a product or service meets or exceeds customers' expectations. However, customers' evaluations are based on their expectations regarding a variety of attributes. Therefore, for firms to improve quality, they must determine the attributes customers use to evaluate their service.

In this chapter we have described some techniques used by service firms to begin to better understand the structures of their services as customers perceive them. In particular, the goal of these various exploratory research techniques is to generate lists of customer needs, the building blocks of the mental structures that consumers envision when they evaluate the quality of a service. We also described two methods for imposing some organization on what is often an enormous list of seemingly trivial comments. Later in the book we will show how this organized list is used to compose the formal surveys that provide statistically valid information about the service and its customers.

NOTES

1. See Paul E. Green, Donald S. Tull, and Gerald Albaum, *Research for Marketing Decisions,* 5th ed. Englewood Cliffs, N.J.: Prentice-Hall, 1988.
2. M. Bitner, B. Booms, and M. Tetreault (1990), "The Service Encounter: Diagnosing Favorable and Unfavorable Incidents." *Journal of Marketing* (January), pp. 71-84.
3. Thomas Reynolds and Jonathon Gutman (1988), "Laddering Theory, Method, Analysis and Interpretation." *Journal of Advertising Research* (January), pp. 11-31.
4. Jonathon Gutman (1982), "A Means-End Chain Model Based on Consumer Categorizaion Processes." *Journal of Marketing* 46 (2), pp. 60-72.
5. Thomas J. Reynolds and Alyce Byrd Craddock (1988), "The Application of the MEC-CAS Model to the Development and Assessment of Advertising Strategy: A Case Study." *Journal of Advertising Research* (April-May), pp. 43-54.
6. Eric von Hippel (1986), "Novel Product Concepts from 'Lead Users'." *Management Science* (July), pp. 791-805.
7. Thomas Peters and Robert Waterman, Jr., *In Search of Excellence.* New York: Harper & Row, 1982.
8. Tom Peters and Nancy Austin, *A Passion for Excellence.* New York: Random House, 1985.
9. Op. cit., p. 23.
10. Abbie Griffin and John R. Hauser (1993), "The Voice of the Customer." *Marketing Science* 12 (Winter), pp. 1-25.
11. See A. Parasuraman, Valarie Zeithaml, and Leonard L. Berry (1988), "SERVQUAL: A Multiple-Item Scale for Measuring Consumer Perceptions of Service Quality." *Journal of Retailing* 64 (1), pp. 12-40. Also, Valarie A. Zeithaml, A. Parasuraman, and Leonard L. Berry, *Delivering Quality Service: Balancing Customer Perceptions and Expectations.* New York: The Free Press, 1990.
12. For example, J. Carman, "Consumer Perceptions of Service Quality: An Assessment of the SERVQUAL Dimensions." *Journal of Retailing* 66 (Spring), pp. 33-55; and J. Paul Peter, Gilbert A. Churchill, Jr., and Tom J. Brown (1992), "Caution in the Use of Difference Scores in Consumer Research." *Journal of Consumer Research* 19 (March), pp. 655-662.

3

THE PSYCHOLOGY OF
CUSTOMER SATISFACTION

PERCEPTIONS AND REALITY

Traditional quality management arose in manufacturing through the pioneering quality-control efforts of Shewhart, Deming, and others. Their focus was on such things as engineering tolerances and defect rates, measured objectively by scientific instruments. The resulting approach to quality was primarily internally focused. If we instead view the organization as a service, then what matters is quality as perceived by the customer. If the customer perceives that quality is bad, then it matters little if "objective" quality is good. Because we are dealing with customer perceptions, the relevant domain is psychology rather than engineering. To measure customer satisfaction effectively, then, we must have some understanding about how satisfaction works.

The disparity between objective quality and perceived quality can be very frustrating to management. For example, customers in the United States have tended in recent years to consider Japanese cars to be high-quality and American cars to be something less. In recent years, Chrysler marketed a car under a domestic nameplate that was actually identical, in all but the most superficial trim, to a car marketed under a Japanese nameplate. Everything about the cars was the same, and they were made in the same plant. Objectively, there was no difference between the two cars. An interesting thing happened, though: customers consistently rated the American version lower! This made Chrysler management tear its hair out. They knew for a fact that their car was just as good, but the quality perceived by customers was lower nevertheless.

Another example of perception versus reality involves the psychology of wait-ing lines. Disney World in Orlando is well known for manipulating customers' minds by playing tricks with its waiting lines. These tricks include having the line make many turns, so no particular segment is very long, and providing entertain-ment along the way so the wait seems shorter. Some psychological principles of waiting lines are:[1] (1) unoccupied time seems longer, (2) preprocess waits seem longer (in other words, it is better to let a patient wait in an examining room instead of the waiting room), (3) anxious waits seem longer, (4) uncertain waits seem longer, (5) unexplained waits seem longer, (6) unfair waits seem longer, and (7) solo waits seem longer.

If *perceived* quality is what matters, then why not deceive the customer? Why not provide inferior quality but persuade the customer that it was great? This issue is illustrated well by the example of a hospital. Let us suppose that Hospital A has excellent technical treatment, including the latest equipment and the most knowl-edgeable doctors, but that it has an unpleasant atmosphere and an unfriendly staff. The customer (patient) cannot easily judge technical quality, but the unpleasant atmosphere and unfriendly staff are easy to identify. Hospital A's perceived quality is thus likely to be poor.

On the other hand, consider Hospital B. Hospital B isn't as good on the techni-cal aspects. Its equipment is older, and its doctors are a little bit behind the times. Yet it will invariably receive very good perceived-quality scores because of its pleasant atmosphere and friendly staff. Should we conclude that Hospital B is a bet-ter hospital because its perceived quality is higher? After all, isn't quality *as per-ceived by the customer* the true indicator of quality?

We can resolve this paradox by considering the element of time. Presumably the patients of Hospital A will have better clinical outcomes, meaning that they will be healthier and happier as time goes by. Their perceived quality rating for Hospital A will tend to reflect their improved health. Hospital B's patients, initially happy, will become less enamored with Hospital B as their health deteriorates over time. Thus, *when viewed in the long term,* perceived quality does tend to converge on objective quality, to the extent that objective aspects are important and to the extent that they become known.

The element of time also becomes important when transactions are repeated over time. In this case, there are two kinds of perceived quality: that of the individ-ual transaction and that of the service provider overall. We can see the difference very easily. Suppose, for example, that you go to dinner at your favorite restaurant, which has been a pleasant experience many times before. However, that particular night the food is bad. The perceived quality of that particular service encounter is negative, but the overall perceived quality of the restaurant is still likely to be good, although perhaps not quite as good as before.

What are the elements of perceived quality? Some authors claimed to have reduced the psychological facets of perceived service quality to a handful of service satisfaction dimensions, which are presumed to be universal across all people. Para-suraman, Zeithaml, and Berry proposed 10 dimensions[2] (later reduced to a smaller number), while Gronroos proposed six.[3] These categorization schemes turned out to be not as universal as the authors had hoped because of the profound differences between service scenarios. Current thought is that no list of universal service quality dimensions exists. Different services are truly different and involve different wants and needs.

SATISFACTION: AN EMOTIONAL RESPONSE

Many managers treat service quality and customer satisfaction as though they were interchangeable concepts.[4] However, experts in customer satisfaction agree that the concepts are quite distinct.[5,6] In particular, perceived quality is a rational perception, while satisfaction is an emotional or feeling reaction.[7] Satisfaction states may include contentment (the phone works), surprise (I won the lottery!), pleasure (the wine is good), or relief (the dentist is finished drilling).[8]

Perceived quality *influences* satisfaction, but it is not the same thing. For example, if a new Rolls-Royce is delivered with a slight nick in the paint, then perceived quality is still likely to be high (higher than just about any other car), but satisfaction will be low. That is because the Rolls-Royce buyer has very high expectations. Absolutely nothing is supposed to be wrong. Consider, on the other hand, the buyer of a used Yugo. That car can be a complete wreck (very low perceived quality), but if it still runs at all, the buyer is likely to be very happy (high satisfaction). That is because expectations are very low.

The word satisfaction comes from the Latin words *satis* (enough) and *facere* (to do or make).[9] These words suggest the true meaning of satisfaction, which is fulfillment. Managerially, fulfillment usually translates to solving problems. Going beyond mere satisfaction involves doing more than eliminating problems; it involves the concept of delight.

DELIGHT

Delight is positive surprise. Although often considered one form of satisfaction, it is the most extreme form, and it relates managerially to better outcomes (higher customer retention levels, etc.) than can be gotten through mere satisfaction.

The business press is full of references to "delighting the customer"; however, there is considerable confusion among managers about what delighting the customer really means. It is often referred to as some sort of complete problem solving, as in "The customer is satisfied with a defect rate of .01 percent, but if we could eliminate defects altogether, then the customer would be delighted!" Actually, at such a small defect rate, it would likely be some time before the customer even noticed. Eliminating problems may satisfy the customer, but delight requires surprise. Things must happen that the customer considers extraordinary. Consider, for example, an airline flight leaving from Nashville. If the plane left on time, then the customers might be satisfied. On the other hand, suppose all families traveling with small children were given toy guitars to take home. That would be a positive (presumably) surprise, and could produce delight.

Delight is possible only if the customer is satisfied to begin with. For example, suppose an airliner had all its engines go out, and was plummeting to earth, out of control. This would make all but the most fatalistic passengers very dissatisfied. Suppose now that the flight attendants went up and down the aisle giving out gourmet chocolates. This positive surprise, which might ordinarily produce delight, would not work in this instance, because the base level of satisfaction had not been established. This sort of hierarchy of satisfaction has been known to academic researchers for years.[10]

Delight results in behavioral outcomes (repurchase, positive word of mouth, etc.) that are substantially better than mere satisfaction can provide. This relationship is "nonlinear."[11,12] In other words, there are thresholds of satisfaction beyond which little benefit is obtained. Referring to Figure 3-1, we see that improvement in

FIGURE 3-1   THE EFFECT OF SATISFACTION AND DELIGHT

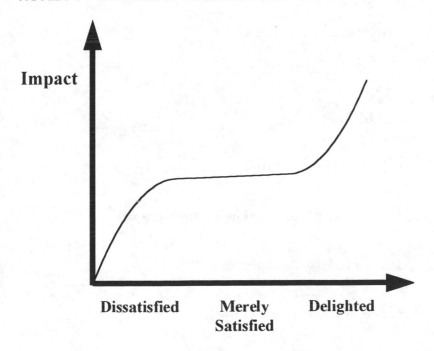

satisfaction first has considerable impact (steep slope of the impact curve) because problems are being solved. Much further progress is not experienced until the extreme right of the curve, the point at which positive surprise (delight) is reflected. Delight is very effective at building loyalty and devotion, much more so than mere satisfaction.[13]

EXPECTATIONS

Satisfaction and delight are both strongly influenced by customer expectations. The term "expectations" has been used by behavioral researchers in a way that is not as precise as the usage by mathematicians, which is "what is likely to happen, on average." Instead, we find a bewildering array of "expectations" that reflect what might, could, will, should, or better not happen (see Figure 3-2).

The *will expectation*[14,15] comes closest to the mathematics definition. It is the average level of quality that is predicted based on all known information. This is the expectation level most often meant by customers (and used by researchers). When someone says that "service exceeded my expectations," what they generally mean is that the service was better than they had predicted it would be.

FIGURE 3-2   THE EXPECTATIONS HIERARCHY

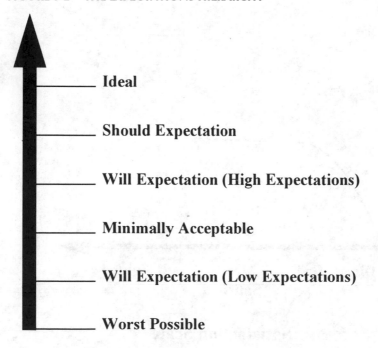

Ideal

Should Expectation

Will Expectation (High Expectations)

Minimally Acceptable

Will Expectation (Low Expectations)

Worst Possible

The *should expectation*[16,17] is what the customer feels he or she deserves from the transaction. Very often what *should* happen is better than what the customer actually thinks *will* happen. For example, a student may think that each lecture *should* be exciting but doubts that that day's lecture actually *will*.

The *ideal expectation*[18] is what would happen under the best of circumstances. It is useful as a barometer of excellence. On the other end of the scale are the *minimally acceptable level*[19,20] (the threshold at which mere satisfaction is achieved) and the *worst possible level* (the worst outcome that can be imagined).

Expectations are strongly affected by experience. For example, if the customer has a bad experience, then the *will expectation* usually declines. A good experience tends to raise the *will expectation*. Generally speaking, the *should expectation* will ratchet up but never decline. Thus expectations change over time, often for the better.

An example of this is the U.S. auto industry. General Motors, Ford, and Chrysler had instilled a level of quality expectations in the American population that was low by today's standards. Then the Japanese started exporting cars of significantly higher quality. Expectations jumped as customers saw that a higher level of quality was possible. The complacent American automakers, making cars of the same quality as always, suddenly found themselves faced with millions of customers with significantly higher expectations. The result was disastrous.

Experience is not the only thing that shapes expectations. Expectations may also be affected by advertising,[21] hearsay,[22] and personal limitations.[23] However expectations are formed, they are a very important influence on satisfaction.

DISCONFIRMATION (GAPS)

The gap between perceived quality and expected quality, called "expectancy disconfirmation," is a very powerful predictor of satisfaction.[24] In fact, this link is so strong that satisfaction itself has often been (incorrectly) *defined* as the gap between perception and expectation.[25] The importance of disconfirmation in explaining satisfaction and other behavior has been demonstrated in many contexts, including sales force interactions,[26] restaurant service,[27] security transactions,[28] and telephone service.[29] Consider, for example, Figure 3-3. Here perceived quality is higher than expected. This will usually result in satisfaction, and will almost always result in the will expectation being raised. Figure 3-4 shows the opposite; perceived quality is not as good as expected. This will probably result in dissatisfaction, and will very likely result in lowered expectations for the service. These disconfirmations (gaps) form the conceptual basis for the SERVQUAL[30] model for service quality and satisfaction. Disconfirmation also forms the basis for the satisfaction measurement approach used by the ROQ model, described later in the book.

FIGURE 3-3   POSITIVE DISCONFIRMATION

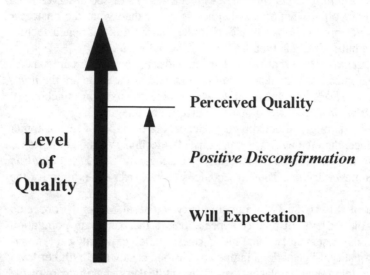

FIGURE 3-4   NEGATIVE DISCONFIRMATION

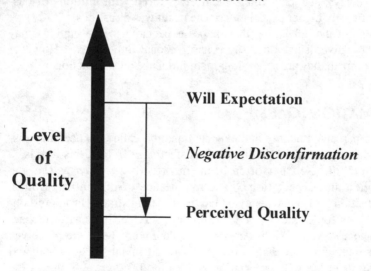

VALUE

Ultimately, it is perceived value that drives purchase and repurchase. Value is formed by the relationship between quality and price. Figure 3-5 shows how quality, price, and value are related. The higher the quality, the higher the value; the higher the price, the lower the value. In colloquial usage, "value" is often used as a

FIGURE 3-5   PRICE, QUALITY, AND VALUE

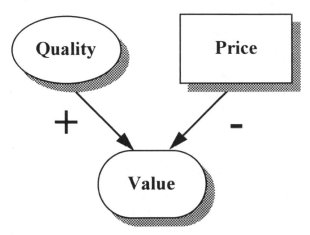

code word for price, and thus really refers to low price in many advertisements. However, use of the word is correct only if quality is constant, which is rarely the case.

Thus we see that there are many ways to obtain value. A product or service may be of relatively low quality, but because it is also very cheap it is a good value. Likewise, a product or service may be very expensive and yet still be a good value because its quality is so high. Ultimately, individual preferences dictate whether there is good value or not.

UTILITY AND CHOICE

The concept of economic utility is a useful way to visualize the relationship between quality, value, and choice.[31] We may think of utility as being some sort of quantifiable "goodness." As quality increases, utility increases (Figure 3-6). As price increases, utility (usually) decreases (equivalently, disutility increases; see Figure 3-7).

Different people may have different utility functions, and this helps explain why different people make such different decisions. For example, consider Figure 3-8. The figure graphs the disutility of price to two people, one rich and one poor. It is easy to see that price A is simply beyond the resources of the poor person, while it incurs very little disutility for the rich person.

Value may be seen as:

$$\text{Value} = \text{Utility of Quality} - \text{Disutility of Price}$$

FIGURE 3-6   THE UTILITY OF QUALITY

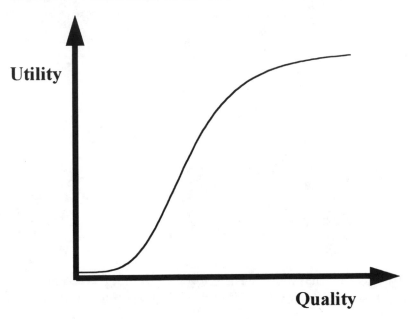

FIGURE 3-7   THE DISUTILITY OF PRICE

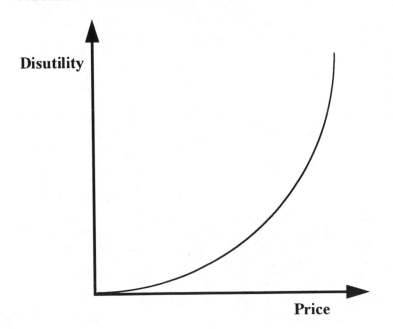

FIGURE 3-8   THE DISUTILITY OF PRICE FOR DIFFERENT INDIVIDUALS

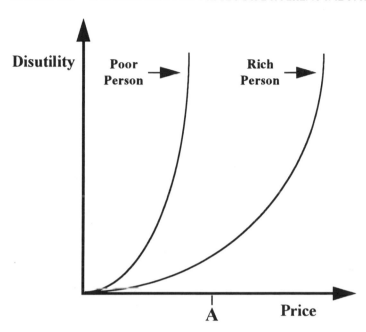

Choice, then, is based primarily on getting the best value. Figure 3-8 shows clearly that there will be market segments in terms of quality. Some people will be happy to pay for more quality, because they will perceive the corresponding increase in price as having little disutility. Others will refuse to pay for more quality, because the price increase has great disutility to them.

Another aspect that becomes important in determining choice is uncertainty. Figure 3-9 shows two persons' expectations of quality outcome. Person 1 is experienced with the product or service, and thus has little doubt about what quality level will ensue. This is illustrated by the fact that the distribution of expected outcomes is tightly bunched. Person 2, on the other hand, is not really sure what will happen because of inexperience. This is illustrated by the broad, unfocused distribution of expected outcomes.

This contrast is important, because there may be downside risk. In other words, people generally find the potential losses from worse-than-expected outcomes to more than outweigh the potential gains from better-than-expected outcomes; so, "worse than expected" hurts more than "better than expected" helps. Thus, if we look at Figure 3-9, even though Person 1 and Person 2 have the same will expectation, Person 1 is most likely more positive than Person 2 because of less downside risk.

FIGURE 3-9   UNCERTAINTY IN EXPECTATIONS

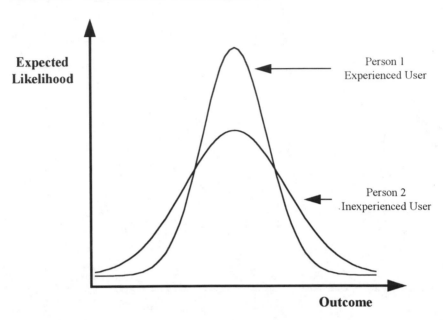

This explains an interesting paradox. Under some circumstances it is perfectly rational for an individual to choose an option that actually is expected to be worse (on average) if the downside risk for that option is less. One thing that tends to reduce uncertainty, and thus worry about downside risk, is experience. As experience increases, knowledge about the product or service increases, and the distribution of expected outcomes tightens up to look more like Person 1 in Figure 3-9. Downside risk is reduced, and probability of repurchase therefore increases, even if the perceived quality is only what was expected. This helps explain why customers often appear loyal. They are being rational and avoiding risk.

THE SATISFACTION PROCESS

Now that we have discussed quality, expectations, disconfirmation, and satisfaction in some detail, it is time to step back and consider how these elements fit together. Figure 3-10 presents a simplified diagram that nevertheless includes the most important linkages. We see that perceived quality results from both objective quality and expectations. Expectations have a direct effect on perceived quality: the higher the expectations, the higher the perceived quality. Perceived quality is then compared to expectations, resulting in a disconfirmation, either positive or negative. Perceived quality also updates the expectations, to produce new expectations

FIGURE 3-10   THE SATISFACTION PROCESS

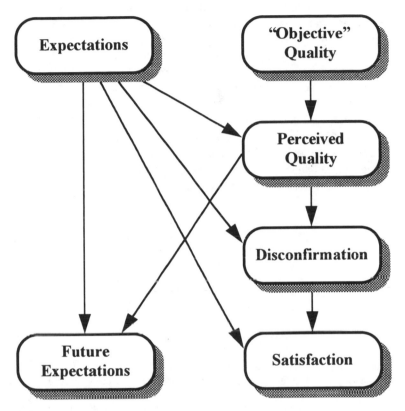

that are either raised (because higher-than-expected quality was experienced) or lowered (because lower-than-expected quality was experienced). At the same time, satisfaction results primarily from disconfirmation, but also from expectations (secondarily). In other words, there is also a direct effect of expectations on satisfaction—the higher the expectations, the higher the satisfaction.

SUMMARY

Although traditional quality management has focused on objective quality, it is customers' perceptions of quality that really count. Therefore, for firms to understand how customers perceive their quality, they must measure customers' satisfaction with their products or services. To do this effectively requires an understanding of the psychology of customer satisfaction. (It is important to remember, however, that perceived quality and customer satisfaction are distinct concepts.)

The most powerful predictor of customer satisfaction is the gap between perceived and expected quality (called disconfirmation). Customers can either experience negative disconfirmation (perceived quality is worse than expected), no disconfirmation (perceived quality is about as expected), or positive disconfirmation (perceived quality is better than expected). Negative disconfirmation tends to be associated with customer dissatisfaction, while no disconfirmation tends to be associated with customer satisfaction. Positive disconfirmation is usually associated with customer delight. It is important to note that customers must be satisfied before they can be delighted. Finally, while perceived quality is an important factor in customers' decisions to purchase or repurchase a service, it is not the only factor. Ultimately, a customer will choose a service based on the utility it provides relative to competition. This means that customers will make tradeoffs between such factors as price and quality when making purchasing decisions.

NOTES

1. David H. Maister (1985), "The Psychology of Waiting Lines," in *The Service Encounter,* John A. Czepiel, Michael R. Solomon, and Carol F. Surprenant, eds. Lexington, MA: Lexington Books, pp. 113-123.
2. A. Parasuraman, Valarie A. Zeithaml, and Leonard L. Berry (1985), "A Conceptual Model of Service Quality and Its Implications for Future Research." *Journal of Marketing* 49 (Fall), pp. 41-50.
3. Christian Gronroos (1988), "Service Quality: The Six Criteria of Good Perceived Service Quality." *Review of Business* 9 (Winter), pp. 10-13.
4. William Boulding, Ajay Kalra, Richard Staelin, and Valarie A. Zeithaml (1993), "A Dynamic Process Model of Service Quality." *Journal of Marketing Research* 30 (February), pp. 7-27.
5. Joseph J. Cronin and Steven A. Taylor (1992), "Measuring Service Quality: A Reexamination and Extension." *Journal of Marketing* 56 (July), pp. 55-68.
6. Richard L. Oliver (1993), "A Conceptual Model of Service Quality and Service Satisfaction: Compatible Goals, Different Concepts," in *Advances in Services Marketing and Management: Research and Practice,* Teresa A. Swartz, David E. Bowen, and Stephen W. Brown, eds. Vol. 2. Greenwich, Conn.: JAI Press.
7. Robert A. Westbrook, Joseph W. Newman, and James R. Taylor (1978), "Satisfaction/Dissatisfaction in the Purchase Decision Process." *Journal of Marketing* 42 (October), pp. 54-60.
8. Richard L. Oliver (1989), "Processing of the Satisfaction Response in Consumption: A Suggested Framework and Research Propositions." *Journal of Consumer Satisfaction, Dissatisfaction, and Complaining Behavior* 2, pp. 1-16.
9. Oliver (1993), op. cit.
10. John E. Swan and Linda Jones Combs (1976), "Product Performance and Consumer Satisfaction: A New Concept." *Journal of Marketing* 40 (2), pp. 25-33.
11. Kevin Coyne (1989), "Beyond Service Fads—Meaningful Strategies for the Real World." *Sloan Management Review* 30 (Summer), pp. 69-76.

12. Terence A. Oliva, Richard L. Oliver, and Jan C. MacMillan (1992), "A Catastrophe Model for Developing Service Satisfaction Strategies." *Journal of Marketing* 58 (July), pp. 83-95.

13. Dave Ulrich (1989), "Tie the Corporate Knot: Gaining Complete Customer Commitment." *Sloan Management Review* 30 (Summer), pp. 19-27.

14. Boulding et al. (1993), op. cit.

15. John A. Miller (1977), "Studying Satisfaction, Modifying Models, Eliciting Expectations, Posing Problems and Making Meaningful Measurement," in *Conceptualization and Measurement of Consumer Satisfaction and Dissatisfaction,* H. Keith Hunt, ed. Cambridge, Mass: Marketing Science Institute, pp. 72-91.

16. Miller (1977), op. cit.

17. Boulding et al. (1993), op. cit.

18. Miller (1977), op. cit.

19. Miller (1977), op. cit.

20. Valarie A. Zeithaml, Leonard L. Berry, and A. Parasuraman (1993), "The Nature and Determinants of Customer Expectations of Service." *Journal of the Academy of Marketing Science* 21 (Winter), pp. 1-12.

21. Raymond P. Fisk and Kenneth A. Coney (1981), "Postchoice Evaluation: An Equity Theory Analysis of Consumer Satisfaction/Dissatisfaction with Service Choices," in *Conceptual and Empirical Contribution to Consumer Satisfaction and Complaining Behavior,* H. Keith Hunt and Ralph L. Day, eds. Bloomington, Ind.: Indiana University.

22. Robert E. Burnkrant and Alain Cousineau (1975), "Informational and Normative Social Influence in Buyer Behavior." *Journal of Consumer Research* 2 (December), pp. 206-215.

23. Westbrook et al. (1978), op. cit.

24. Richard L. Oliver (1980), "A Cognitive Model of the Antecedents and Consequences of Satisfaction Decisions." *Journal of Marketing Research* 17 (November), pp. 460-469.

25. John A. Howard and Jagdish N. Sheth, *The Theory of Buyer Behavior.* New York: John Wiley & Sons, Inc., 1969.

26. Richard L. Oliver and John E. Swan (1989), "Equity and Disconfirmation Perceptions as Influences on Merchant and Product Satisfaction." *Journal of Consumer Research* 16 (December), pp. 372-383.

27. John E. Swan and I.F. Trawick (1981), "Disconfirmation of Expectations and Satisfaction with a Retail Service." *Journal of Retailing* 57, pp. 49-67.

28. Richard L. Oliver and Wayne S. DeSarbo (1988), "Response Determinants in Satisfaction Judgments." *Journal of Consumer Research* 14 (March), pp. 495-507.

29. Ruth Bolton and James Drew (1991), "A Multistage Model of Customers' Assessments of Service Quality and Value." *Journal of Consumer Research* 17 (March), pp. 375-384.

30. A. Parasuraman, Valarie A. Zeithaml, and Leonard L. Berry (1988), "SERVQUAL: A Multiple-Item Scale for Measuring Consumer Perceptions of Service Quality." *Journal of Retailing* 64 (1), pp. 12-40.

31. William D. Perreault, Jr. and Frederick A. Russ (1976), "Physical Distribution Service in Industrial Purchase Decisions." *Journal of Marketing* 40 (April), pp. 3-10.

4

DESIGNING CUSTOMER SATISFACTION SURVEYS

OVERVIEW

Judging from the number of bad customer satisfaction surveys, it would seem that just about everyone thinks that they are capable of writing a customer satisfaction survey with little or no training. Unfortunately, that is incorrect, as the large number of surveys that produce unactionable results would attest. It is surprising how many surveys—even those designed and written by major marketing research firms and administered by large corporations—are fundamentally flawed and essentially worthless. Nevertheless, by following some basic principles, it is fairly easy for even small companies to design a very useful survey. Designing a good customer satisfaction survey is a skill, but one that can readily be acquired through diligent study and practice.

This chapter presents a systematic approach to designing and writing a customer satisfaction survey. This approach is not the only one that could be adopted, but it is a proven, tested approach that has been used successfully for hundreds (if not thousands) of surveys by a variety of organizations, including some of the most advanced companies in the world.

Figure 4-1 shows, in summary form, the steps in designing and implementing a customer satisfaction survey. One must begin with essential preliminaries, such as obtaining top management buy-in and conducting exploratory research. From there, the sampling and data collection details should be worked out and the key business processes identified. The emphasis on making the survey relevant to business processes is the foundation of our approach and the point at which most surveys fail most completely.

FIGURE 4-1   THE CUSTOMER SATISFACTION SURVEY PROCESS

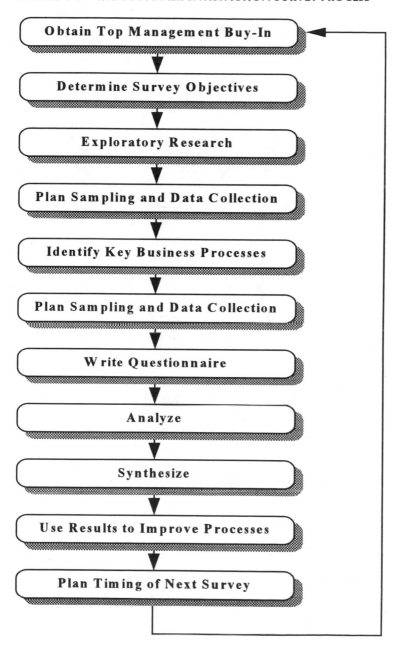

The survey is written according to a structure dictated by the business processes, the results are analyzed, and the main findings are synthesized. The main findings are then used to improve the business processes, and subsequent surveys are planned to monitor progress and identify areas of further opportunity.

PRELIMINARIES

The first thing to figure out is why the customer satisfaction survey is being conducted in the first place. Answers such as "to listen to the customer" or "because everybody has one" indicate a lack of clear motivation. Good reasons for doing a customer satisfaction survey include:

1. To find out where to focus process improvement efforts;
2. To determine whether previous improvement efforts have worked; or
3. To see whether strategic advantages or disadvantages exist.

It is imperative that the reason for doing the survey be clearly stated and understood ahead of time.

Little can be done with customer satisfaction survey results without the approval of top management. Thus, it is essential that the survey project have top management buy-in. Top management should agree with the purpose of the research and understand what actions will result from the survey. One way of ensuring management buy-in is to involve management in the design and construction of the survey. Too much top management meddling is not good, but it is helpful to make sure that management sees drafts of the questionnaire and has the opportunity to suggest changes. This prevents the research from getting blindsided after the fact by a manager who says something equivalent to, "This question is stupid; you would have gotten different results if you had asked the question right."

It is also worth remembering that in any service, it is the front-line employees who have the bulk of the customer contact and thus the greatest opportunity to implement process improvements that directly affect the customer. Because the front-line employees are the ones who will make most of the important changes, it is essential that their voice be represented in the construction of the survey. They should have input all the way through and should be represented on any committees entrusted with designing the survey and implementing its results.

Involving both management and front-line employees is crucial, because it lays the groundwork for effective implementation of the survey's findings. Advance effort is also needed to ensure that the survey is relevant to customers. One of the most common errors in designing a customer satisfaction survey is to simply write the survey "cold," without adequately exploring the wants and needs of the customer. This may appear to save time and money, but in the long run it is actually a foolish waste of energy, because the survey is sure to be worthless.

It is absolutely essential to conduct an exploratory research phase prior to writing the questionnaire. This phase ensures that the issues covered by the questionnaire are relevant to the customer and that important customer concerns are on the questionnaire. Many techniques that are useful for this purpose are discussed in Chapter 2. In general, the idea is to structure the questionnaire around business processes. Issues concerning the subprocesses (dimensions) within each process should be those that are "top of mind" to customers. Those issues should be in the customers' own words, as derived from focus group transcripts, complaints, and suggestions.

SAMPLING

Once the preliminaries are out of the way, it is time to consider the important issues of what the sample population should be and how the sample should be drawn. The sample population, depending on the purpose of the survey, could include an organization's current customers, former customers, competitors' customers, or prospective customers.

Current customers are a good place to start. They are familiar with the service provided and also represent potential future profits through repurchase and referrals. It is actually more difficult to sample a firm's current customers than one might think. Let us consider some of the possibilities. First, suppose we are a retail bank. Our customers are well-known to us. We have all sorts of information about them, including name, address, phone number, and account information. It is very easy to contact them.

On the other hand, suppose we are a summer boat-rental business. Who are our current customers, especially if it is winter? We might suspect that a customer who has returned for several consecutive summers is a current customer, but we do not know for sure. That same customer may have moved away, or died. These customers do not come in to "close their boat account" when they decide they don't want to do business with us anymore. They just stop coming back. In such a case it is best to sample customers as they conduct their transaction, perhaps contacting them two or three days after they have returned home. Alternatively, we may be able to administer the questionnaire as they prepare to leave—perhaps as part of the payment process, if the questionnaire is short enough.

Former customers are also good to talk to. Presumably we did something wrong to cause them not to return. If we sample only current customers and not former customers as well, we may get too rosy a picture of how we are doing. One way to question former customers is as they leave. For example, when a customer closes a bank account, the bank can ask some satisfaction questions, perhaps even why the customer is leaving. If the business has customer addresses or phone numbers, it can also contact them after they have left. Businesses that conduct business on a transaction basis but do not keep detailed customer information, such as a restaurant, have difficulty determining former customers. In such cases, the only recourse is to sample the general population, or some more targeted version of the general population, and screen out the former customers. However, this is generally too expensive to be practical.

Talking to competitors' customers is also a good idea, if the budget permits. It is very useful strategically to know where the competition is strong and where it is weak. One method often used is sampling the competitor's customers as well as one's own customers, administering exactly the same questionnaire. However, comparisons between the two groups are often made that lead to incorrect conclusions. Suppose, for example, that satisfaction with reliability of phone service averages 3.5 (on a 5-point satisfaction scale, with 5 best) for our customers and 2.5 for the competitor's customers. One might conclude that we are perceived as better on reliability. Taking this further, one might subtract the scores and say that we had a 1-point advantage on reliability. The above analysis is commonly employed to evaluate performance relative to the competition. You can see why this can be wrong by considering some illustrative examples.

Let us suppose, for example, that our customers are predominantly American, and their customers are predominantly Japanese. Suppose also that Japanese are pickier, on average. In such a case our competitor's 2.5 may be just as good or better than our 3.5. We may be misled if we think we are better.

Few people would make the mistake of comparing customer satisfaction levels across different countries without making some sort of adjustment for national differences. But what about within a country? Suppose, for example, that our customers are predominantly from the South, and the competitor's customers are predominantly from the North. Suppose Northerners are pickier. Again, our apparent advantage may be illusory; one must adjust the ratings by considering regional variation.

Even if we adjust for all geographic differences, we still may have bias in our comparison. Consider, for example, the case in which our customers are older, and older people are less picky. In such a case a bias will again exist. The comparison is no good unless age differences are considered.

You can see where this is going. Actually, any customer characteristics that may be related to the tendency to rate high or low must be considered when making a comparison with competition. Their customers are not the same as ours, and we should not pretend that they are if we wish to make a valid comparison with the competition.

Another group of customers to question is prospective customers. Although they will not be able to respond to satisfaction questions, because satisfaction requires experience, they will be able to respond to questions about perceived quality of our firm and competitors. This is useful in evaluating offensive marketing opportunities and weaknesses. Why are they not our customers already? Where do we need to strengthen our image?

Regardless of which group you survey, it is important that the sample be drawn in a scientific manner. In particular, it is important that the sample be drawn in such a way that a "probability sample" results. A probability sample is one in which any member of the population has a known (at least in principle) probability of being selected. Strictly speaking, it is not correct to do statistical analysis on samples drawn in any other way.

The simplest way to ensure a probability sample is to generate a complete list of the population, and then use random numbers to select the sample. Such a sample is known as a *simple random sample.* An even simpler alternative is to take a random starting place on the list, and then sample every nth person (every 10th, every 500th, or whatever is necessary to draw a sample of the size required). This is known as a *systematic random sample.* Unless there is some sort of recurring pattern in the list, this method will result in samples that are almost as good as simple random samples. Other probability sampling techniques, such as stratified sampling and cluster sampling, are too involved to discuss here, but may be studied in any marketing research text.

One sampling method seems scientific but does not result in a probability sample. That method is *quota sampling.* To collect a quota sample, the researcher tries to categorize people into groups defined by classifier variables, such as age and sex. Then, based on the percentage of people in the population that should be in each age-sex category, the number to be surveyed in each age-sex cell is obtained. Then the researcher goes around and tries to hunt down people who would fit in the cells. Because this hunting process is rarely random (how many of you have seen interviewers in shopping malls trying to chase down a man 60–69 years old to complete their quota?), some individuals inevitably have a higher probability of being selected. For example, someone who looks dangerous, different, or just in a hurry will almost certainly be skipped. Quota sampling should be avoided if at all possible.

DATA COLLECTION

How the survey data are collected can have a strong influence on their quality and usability. The first major decision that must be made is that of the *mode of data collection,* or what technical means is used to collect the data.

Telephone surveys are popular in customer satisfaction measurement for several reasons. The most important reason is that response rates tend to be good. A response rate of 80 percent or better is not unusual, if the survey is interesting and well designed. Telephone interviewing is also very fast, because electrical impulses travel at the speed of light. A typical setup is for the telephone interviewer to be seated at a computer or terminal, with the questionnaire projected on the screen. The keypad is then used to log the responses, which are then automatically fed into a database. This makes it possible to analyze the data almost immediately, because data entry is incorporated directly into the interviewing process. One disadvantage of telephone interviewing can be cost, because interview time is not cheap and long distance phone bills can add up.

If an individual on the sample list is not reached on the first call, it is important to call back later, perhaps several times and at different times of the day, to try to obtain his or her response. This helps to maintain the integrity of the sample. Otherwise, the sample becomes biased in favor of people who are home a lot, such as retirees and housewives, and biased against those who are rarely home, such as businesspeople and two-career households.

Mail surveys are also popular, especially because they are relatively inexpensive. The main argument against mail surveys is low response rates. The naive researcher often concludes that this is not a problem. For example, if 200 completed surveys are needed, and a 20 percent response rate is expected, then 1,000 surveys can be sent out. However, this reasoning is incorrect, because the real problem with a low response rate is not the small number of eventual respondents, but rather nonresponse bias. If only 20 percent of the sample responds, which 20 percent is it? Is it the least busy 20 percent, the most interested 20 percent, or what? The respondents may differ from the population in important ways, which seriously damages the validity of the survey results. On the other hand, if 90 percent of the sample responds, then there is not much opportunity for that 90 percent to be unusual, because the respondents match the sample (and thus, by inference, the population) almost completely. It is revealing that this potential for nonresponse bias is such a problem that some major marketing research firms refuse to conduct mail surveys.

Nevertheless, there are occasions in which budget constraints dictate that a mail survey be employed. In those cases, it is imperative that everything possible be done to increase the response rate. Some of the ways this can be done include keeping the questionnaire short, enclosing a postage-paid return envelope, enclosing a cover letter that motivates the respondent to respond, sending prior notification the week before and reminders the week after, enclosing some sort of incentive, making the survey easy to fill out, and making sure the mailing list is up-to-date.

Personal interviews are also possible. Sometimes interviews are the only feasible way to collect data. For example, how else could one sample the customers of a fast-food restaurant? The only feasible alternative would be to hand out surveys to be filled out later. Our experience is that handout surveys generate *extremely* low response rates, sometimes in the 5 percent range. No one wants to walk around with a survey, so they throw it out the first chance they get. Personal interviews generally have a good response rate, but the big concern is maintaining a probability sample. The way to do that is to have very careful rules about exactly who will be sampled. Using the fast-food example, one might sample every 10th customer through the door, for example. That would make sure that the respondents selected were not just the friendliest, best dressed, and most attractive.

One popular method of collecting data that should definitely be frowned upon is comment cards. These are often seen in hotel rooms, on restaurant tables, and in shops. There are several fatal problems with using comment cards to collect customer satisfaction survey data. First, there is no attention to sampling. In effect the entire population is being halfheartedly sampled. The result is an extremely low response rate. That alone would instill a terrible nonresponse bias problem, but the problem is even worse than it appears, because the people who tend to seek out comment cards are the ones who have extremes of opinion. Either they are ecstatic (occasionally) or they are furious (almost always). Thus the respondent pool is drastically biased in favor of hotheads and grumps. Actually, there is an appropriate use of comment cards, and that is as an exploratory device to generate hypotheses about areas of concern. These hypotheses can then be implemented in customer satisfaction surveys (which are appropriately designed to sample properly and minimize nonresponse bias).

Once you have decided how to administer the questionnaire, the question remains as to who should administer it. It is tempting to reduce cost by administering the survey in-house. This rarely works. Let us consider some examples to understand why. Suppose we are conducting a sample of employee satisfaction, and we decide to save money by having management conduct the survey with personal interviews. Clearly, in such a case the respondents will be intimidated and unable to give honest responses. For another example, suppose the customers are to be given questionnaires by the sales force. Presumably these questionnaires will have questions that reflect (positively or negatively) on the sales force. Under those conditions, the salespeople will be tempted to give the questionnaires only to the customers who are happy, or to fill some of the questionnaires out themselves, just to be safe. In both cases the administration of the questionnaires causes bias, because the people administering the questionnaire are not disinterested parties. The easy way to overcome these problems is to commission an outside researcher or company to administer the questionnaire. Generally this will cost more, but the decrease in bias and the increase in professionalism will usually make this tradeoff worthwhile.

Wording

The wording of the items in a customer satisfaction questionnaire is crucial to the questionnaire's success, and there are many nuances that are routinely butchered by naive questionnaire writers. The first thing to realize is that the questionnaire must be written in language that is relevant to the customer. This is accomplished best by using the words of the customers themselves. One way to do this is to take transcripts of focus group interviews. If several focus group participants talk about something in the same way, then that way is probably the customers' vernacular.

It is easy to forget that company employees develop their own jargon, which differs from common language usage. For example, bankers may talk about "ATM machines," while customers may talk about "cash machines" or "bank machines." Computer people may talk about "bits, bytes, RAM, and ROM," while customers may talk about whether the computer "is big enough to hold my programs." The gap in language can be huge, and the questionnaire must reflect the language of the customer, not the jargon of the company.

One innovative way to discover the customers' language is to put all of the customer responses verbatim into a computer database. The database can then be used to count which words are used the most and how words are used together. Obviously, a computer must be used to perform this onerous counting task.

Once you are confident that you know how customers talk about things, the next step is to decide what mix of closed-end (multiple-choice) and open-end (essay) questions to use. You know from school that multiple-choice tests are faster and easier; the same thing holds for customers. If the questionnaire is full of open-ended questions, it looks lengthy and difficult. This generally means that the number of open-ended questions should be severely limited. Sometimes there is no way of avoiding an open-ended question, but often such a question can be converted into a closed-end. For example, the question, "At what kind of store did you buy the computer?" could be changed to "Where did you buy the computer?: (a) computer store, (b) discount electronic store, (c) mail order, (d) phone or (e) other."

Usually, a great deal can be learned from one or two strategically worded open-ended questions at the end of the questionnaire. The purpose of these questions is exploratory in nature, which means the questions might be marked "optional." One very useful question is, "What is the single biggest problem you have had as a customer?" Another is, "What is the single most important thing we could do to improve our service?" Either question will provide a wealth of information. The verbatim responses from just one of these questions will keep a manager busy for hours, and deeply enrich that manager's intuitive understanding of the most important customer issues. We recommend making the responses available unedited and uncensored.

The closed-ended questions are generally of a standard form in a customer satisfaction survey. Every question is worded roughly the same, with only the subject of the question changing. Typically the goal is to measure the level of performance, whether that performance is quality, satisfaction, or comparison with expectations. Figure 4-2 shows some of the main ways in which the satisfaction question can be worded. The first two examples in the figure are quality questions. Researchers designed the first one to measure the reputation of the company overall. This measure should not change much over time, although it may slowly drift up or down. The second measures quality of a particular transaction. It is important that the researcher know whether the overall reputation of the company or the quality of specific transactions is sought. Generally, it is preferable to measure the quality of transactions, because quality improvement (or deterioration) will be detected quicker by a transaction-specific question. A reputation question involves not only recent transactions, but also to some extent all past transactions.

FIGURE 4-2   WORDING THE SATISFACTION QUESTION

Quality (Cumulative)
Please rate the quality of this restaurant:
☐ Excellent   ☐ Good   ☐ Fair   ☐ Poor

Quality (Transaction - Specific)
Please rate this visit to the restaurant:
☐ Excellent   ☐ Good   ☐ Fair   ☐ Poor

Satisfaction
How satisfied were you with this visit to the restaurant?
☐ Very Satisfied   ☐ Satisfied   ☐ Dissatisfied   ☐ Very Dissatisfied

Disconfirmation
Please rate this visit to the restaurant:
☐ Much Better   ☐ About as   ☐ Worse than
  than Expected   Expected   Expected

The next example is a satisfaction question. Note that satisfaction, by its very nature, is transaction-specific. It also implies experience, whereas quality questions can measure perceptions of quality even without experience. We show a scale from "Very Satisfied" to "Very Dissatisfied," but other scales are possible, such as, for example, a "Delighted" to "Terrible" scale.[1] The last example is a disconfirmation question, measuring comparison with expectations. This last question is the one we advocate.[2] This is because, under some fairly reasonable assumptions, it can be shown mathematically that comparison with expectations will correlate higher with customer retention than either a quality question or a satisfaction question.[3]

Note also the way the categories of this question are set up. It is well known among customer satisfaction researchers that the typical customer satisfaction or quality question produces very skewed responses, with an inordinate number of responses at the very top end of the scale.[4] Essentially, too many people say "Excellent" or "Very Satisfied" when they really just mean that they didn't experience any problems. By making the top category "Much Better than Expected," it is much more difficult for the respondent to check the top box. We have found that the skewness of the scale decreases dramatically, which increases the information imparted by the scale and ensures that the people checking the top box really are delighted. Although this question is a comparison with expectations question, and not a satisfaction question, for convenience we will generally refer to the "Worse than Expected" respondents as "Dissatisfied," the "About as Expected" respondents as "Satisfied," and the "Much Better than Expected" respondents as "Delighted." Because of the very strong link between disconfirmation and satisfaction, this relabeling is not too much of a stretch.

One key issue is how many scale points to use. Good results may be obtained from just about any number of scale points. We recommend using three scale points, in association with a disconfirmation scale, because this enables us to model the effects of satisfaction and delight separately.[5]

While we advocate using a disconfirmation scale when conducting customer satisfaction research, we recognize that some organizations will not adopt this scale. Many firms have already invested in satisfaction measurement systems that use different scales, and they will be reluctant to change. In these cases, we recommend that the data be recoded to a three-point scale for ROQ analysis. Although converting from one scale to another is a crude approximation at best, converting the top score to "delighted," the mid-point to the second highest to "satisfied," and all ratings below the mid-point to "dissatisfied" may produce acceptable results. Using this method for a five-point scale, where 5 is "best" and 1 is "worst," a 5 would be "delighted," 3 and 4 would be "satisfied," and 1 and 2 would be "dissatisfied." While this recoding method does not create a perfect match between scales, it may produce a reasonable approximation.

Unnecessary Questions

I PREFER A 6 POINT SCALE

Some questions frequently appear in customer satisfaction surveys but are not really necessary. These include expectations questions and importance questions. Figure 4-3 shows examples of each. The idea behind the expectations question is to measure quality performance versus expectations.[6] Unfortunately, there are statistical and methodological problems that argue against this practice.[7] Thus the suggested remedy is to measure disconfirmation directly.[8] This has the double advan-

FIGURE 4-3   EXAMPLES OF UNNECESSARY QUESTIONS

Expectations Questions
Please rate the level of quality you expected:

0   10

Very   Excellent
Poor

Importance Questions
How important is airline safety to you?

Very Important	Somewhat Important	Neutral	Somewhat Unimportant	Very Unimportant
☐	☐	☐	☐	☐

tage of being more statistically reliable and cutting the length of the questionnaire in half.

The idea behind including importance questions, such as the one shown in Figure 4-3, is that it is best to concentrate on the issues that are important to customers. Unfortunately, importance questions not only double the length of the questionnaire, they also do not work. Take, for example, the question in Figure 4-3. It is hard to imagine a customer who will not report that airline safety is "Very Important." After all, no one wants to be picked up in charred pieces from the ground and placed in a body bag. At the same time, airline safety is not determinant of either initial choice or satisfaction. Because just about all airlines have indistinguishable safety records (except for commuter airlines), airline safety will hardly ever lead to any important

FIGURE 4-4   QUESTIONNAIRE STRUCTURE

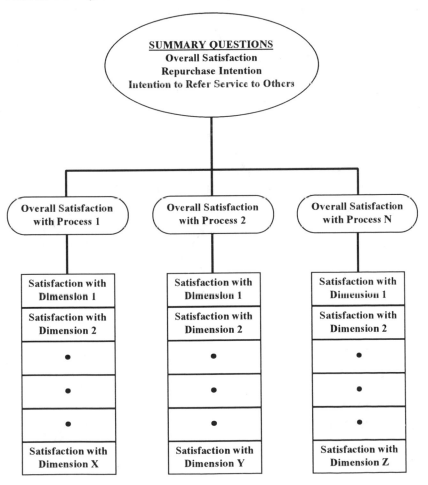

behaviors. For our purposes, airline safety can be ignored. This is not to say that the maintenance crews should be laid off, but rather that the status quo is appropriate, because an incremental shift in safety measures would have little effect on anything measurably important. Determining which dimensions of quality have an impact is still important, of course, but it turns out that this can be done very efficiently using statistical methods on only the satisfaction questions (see Chapter 9).

QUESTIONNAIRE STRUCTURE

Perhaps the most important element of effective questionnaire design (and perhaps the element most frequently done badly) is structuring the questionnaire around business processes. Figure 4-4 shows a typical questionnaire structure. At the top are summary questions. These are linked to overall satisfaction questions on each of the processes. The overall process satisfactions are then linked to satisfaction on dimensions within the process. The reason for building a structure of this sort is that the context of customer satisfaction measurement is quality improvement, and improving quality demands ownership, which is possible only if the results of the survey are linked to business processes. Thus we *structure* the questionnaire around the structure of the business, but within that framework we consider the issues the customer considers relevant, and we word the questions in the customer's language.

Ultimately, we wish to create overall attitudes and behavioral intentions toward our business. Thus we include summary questions on the questionnaire (see Figure 4-5) that will have the closest relationship with actual behavior. The first example question in Figure 4-5 is an overall comparison with actual behavior and with expectations. This will be an overall question, regardless of whether quality, satisfaction, or comparison with expectations is being measured.

FIGURE 4-5   EXAMPLES OF SUMMARY QUESTIONS

Overall Satisfaction/Quality/Disconfirmation
Please rate this visit to the dentist:

☐ Much Better   ☐ About as   ☐ Worse than
   than Expected   Expected   Expected

Repurchase Intention
How likely are you to visit this dentist, the next time you need dental care?

☐ 100%   ☐ 80%   ☐ 60%   ☐ 40%   ☐ 20%   ☐ 0%

Intention to Refer Service to Others
How likely are you to refer this dentist to others?

☐ Very   ☐ Somewhat   ☐ Neither   ☐ Somewhat   ☐ Very
   Likely   Likely   Likely   Unlikely   Unlikely
                                          nor
                                          Unlikely

The second summary question is a repurchase intention question. In many cases, this question can be calibrated against actual repurchase behavior by tracking the respondents over time to check whether they actually repurchase or not. For example, suppose 90 percent of the people who check "100% repurchase intention" actually repurchase. This calibrates the meaning of the "100%" box at 90 percent. Note that this question involves several categories. Another common way of wording the repurchase intention is as a yes-no question. Statistically, it is possible to show that the information contained in a yes-no question is much less, which means that a much larger sample size is required to obtain a similar amount of information. For this reason it is always better to use several categories.

The third summary question is a word-of-mouth question. The idea is to see whether the service was good enough to generate positive word of mouth, which can produce secondary financial benefits through the attraction of new customers. It is difficult to calibrate this question, which limits the question's usefulness.

Following the summary questions are questions about individual business processes (see Figure 4-6). These include overall questions for the processes and detailed questions about their various dimensions. These various questions form a chain that can be used to explain overall satisfaction, repurchase intention, and behavior. The idea is that satisfaction on the process dimensions drives overall satisfaction with the process. The overall process satisfactions then drive overall satisfaction, which drives repurchase intention, which drives repurchase behavior.[9]

Finally, the questionnaire generally includes classification questions, which are used to segment the respondents. These may address a wide variety of possible topics, including demographics, geography, usage occasion, type of customer, or other questions relevant to the managerial scenario being inspected.

FIGURE 4-6   EXAMPLES OF PROCESS QUESTIONS

Processs Overall Questions
Please rate the in-flight meal you were served:
☐ Much Better   ☐ About as   ☐ Worse than
   than Expected   Expected   Expected

Process Dimension Questions
Please rate the vegetables in your in-flight meal:
☐ Much Better   ☐ About as   ☐ Worse than
   than Expected   Expected   Expected

Please rate the dessert in your in-flight meal:
☐ Much Better   ☐ About as   ☐ Worse than
   than Expected   Expected   Expected

OTHER ISSUES

There are a number of other issues that impact the effectiveness of the customer satisfaction measurement process. One issue is questionnaire length. It is certainly the case that too long a questionnaire hurts the response rate. How long is too long? As a rough rule of thumb, any questionnaire longer than about two pages is asking for trouble. We have used questionnaires as short as 4–10 questions with very good results. It is better to start with what is absolutely necessary—and then stop! The enemy of the effective questionnaire is the "nice to know" question, such as, "As long as we're doing a survey, it would be nice to know whether they've seen our advertising or not." "Nice to know" questions are generally mildly interesting to managers but have no impact on management.

A second issue is how often to survey. Often this is dictated by budget. However, companies should realize that surveys are a primary source of market information. Firms that choose to survey infrequently because they focus solely on costs could find themselves left out should changes occur in their markets. A major telecommunications corporation made this very mistake and was blind-sided by its competitors. Major changes had occurred in the market, but they were not detected until months later, when the next survey results were available. As a result, the firm had the unfortunate experience of watching a large number of its customers switch to competitors. As this example illustrates, firms must regularly survey their markets so that they do not lose touch with their customers.

Companies that begin a customer satisfaction measurement program should be very careful about changing the survey. One of the key benefits of customer satisfaction monitoring is to see whether there is improvement (or deterioration) over time. It is difficult to compare results if the questions are always changing, or if the scale is changed. To the extent possible, it is advisable to maintain continuity in the wording of the questions. Likewise, it is advisable to maintain the same scale. If the wording or scale need to be changed, then it is important to do a split test, using each version at the same time for the same population, to see how the old and new versions map into each other.

Many companies are using customer satisfaction scores as one factor in the compensation of employees. This is a proven way to make employees care about customer satisfaction, but great care must be taken to ensure that the measures are used fairly. For example, suppose Salesperson A has a bad bonus this quarter, because factors outside of his or her control caused the customer satisfaction ratings to drop. Suppose, for example, that the shipments to Salesperson A's region were late. In this case, the salesperson will have a great deal of resentment toward the satisfaction measurement program, and probably will do anything possible to rig it or sabotage it in the future.

Or suppose, similar to the problem of comparing satisfaction with competitors, Salesperson A's customers are simply more difficult to please because of some personal characteristic that is prevalent in A's region. A's scores will be poor, all other things being equal, and it is really outside of A's control. Again, Salesperson A will feel as though the system is not fair and will fight it every chance he or she gets. Plans that link satisfaction scores with compensation must be very careful about adjusting for factors that make the comparison unfair.

To get honest responses, especially when sampling employees, suppliers, or channel customers, it is often necessary to promise anonymity or confidentiality. These two things are not the same. *Anonymity* means that the researcher will not know who the respondent is. Thus, tricks such as coding each "anonymous" mail questionnaire with an identifying number are unethical and could lead to legal action against the company. *Confidentiality* means that the researcher knows who the respondent is, but won't tell. This is feasible when the questionnaire is being administered by a third party. It should be made clear to the respondent whether the responses are anonymous or confidential, and then the researcher should scrupulously stick to the bargain. Generally speaking, a promise of confidentiality (or, even better, anonymity) will increase the response rate and make the responses more candid.

Another ethical issue arises with the use of the questionnaire for other than its stated purposes. For example, suppose the real reason the questionnaire is being given is to qualify sales leads. The conversation might go like this:

Interviewer (INT): How satisfied are you with the performance of your vacuum cleaner?

Respondent (RESP): It's OK, I guess.

INT: How long have you owned it?

RESP: About 10 years.

INT: Were you aware that the average life of a vacuum cleaner is 5.7 years?

RESP: Why, no.

INT: Let me tell you about our new vacuum cleaner. . . .

Such use of a customer satisfaction survey is a gross violation of ethics and due cause for investigation by consumer protection agencies.

Another violation is the use of a customer satisfaction survey to impart information about available services. A dialogue might be:

INT: How would you rate your life insurance policy on a 1 to 5 scale?

RESP: Five, I think.

INT: How would you rate your awareness of our homeowners' policies?

RESP: Gee, I didn't know you had homeowners'. One, I guess.

INT: How would you rate your awareness of our car insurance policies?

RESP: You have car insurance, too? I didn't know that. One?

INT: How would you rate your awareness of our personal liability policies? . . .

With more and more "questionnaires" like these, it is no wonder that compliance rates are dropping like a rock.

SUMMARY

In summary, we will emphasize those aspects of designing a customer satisfaction survey that are most often done incorrectly. Key points to remember are:

1. Always do exploratory research before writing a questionnaire.
2. Make sure you have a probability sample.
3. Minimize nonresponse bias. Use phone surveys if possible.
4. Have an objective third party administer the survey.
5. Use transaction questions rather than cumulative questions.
6. Minimize the number of open-ended questions.
7. Use a three-point disconfirmation scale, if possible.
8. Word the scale in such a way that top usage is minimized.
9. Don't use separate expectations questions or importance questions.
10. Structure the questionnaire around business processes.

The next chapter shows how to analyze the results of the questionnaire design discussed in this chapter.

NOTES

1. Robert A. Westbrook (1980), "A Rating Scale for Measuring Product/Service Satisfaction." *Journal of Marketing* 44 (Fall), pp. 68-72.
2. A similar form is recommended in Carman, James M. (1990), "Consumer Perception of Service Quality: An Assessment of the SERVQUAL Dimensions." *Journal of Retailing* 66 (Spring), pp. 33-56.
3. Roland T. Rust and Anthony J. Zahorik (1992), "A Bayesian Model of Quality and Customer Retention," presented at the ORSA/TIMS Joint National Conference, San Francisco.

4. Robert A. Peterson and William R. Wilson (1992), "Measuring Customer Satisfaction: Fact and Artifact." *Journal of the Academy of Marketing Science* 20 (Winter), pp. 61-71.
5. Wayne DeSarbo, Lenard Huff, Marcelo M. Rolandelli, and Jungwhan Choi (1994), in *Service Quality: New Directions in Theory and Practice,* Roland T. Rust and Richard L. Oliver, eds. Newbury Park, Calif.: Sage Publications, pp. 199-220.
6. Parasuraman, Zeithaml, and Berry (1988), op. cit.
7. See Carman (1990), op. cit. and Gilbert A. Churchill, Jr., J. Paul Peter, and Tom Brown (1993), "Caution in the Use of Difference Scores in Consumer Research." *Journal of Consumer Research,* pp. 655-662.
8. Carman (1990), op. cit.
9. Raymond Kordupleski, Roland T. Rust, and Anthony J. Zahorik (1993), "Why Improving Quality Doesn't Improve Quality (Or Whatever Happened to Marketing?)" *California Management Review* 35 (Spring), pp. 82-95.

5

ANALYZING CUSTOMER SATISFACTION SURVEYS

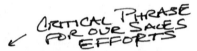

When it comes to market research, the typical manager is not interested in statistics. The manager is interested in knowing "where should I focus my efforts?" Thus, analysis of customer satisfaction surveys should provide results that place possible quality-improvement efforts in some order of priority. This, however, is better accomplished through statistical means than through direct questioning of customers. Also, the "squeaky wheel" method, responding to problems or opportunities according to the number of times the issue is mentioned, fails to identify the most important issues.[1]

Uncovering information contained in customer satisfaction research requires researchers to employ statistical methods to analyze the data. For the information to be useful, however, the results must be presented in a manner that allows managers to easily determine the importance of various quality-improvement alternatives.

OVERVIEW OF PROCESS

Structuring the questionnaire according to business processes, as illustrated in Chapter 4, makes it possible to structure the data analysis in a managerially meaningful way. The analysis determines links between the following components:

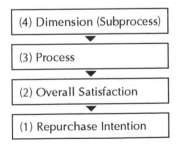

Customers' satisfaction level with the dimensions (subprocesses) of a process affects their satisfaction level with that process. Likewise, customers' satisfaction level with the various processes influences their satisfaction level with the firm overall. Finally, customers' satisfaction level with the firm overall affects their likelihood of repurchasing the service.

An example should make this clear. A hotel manager may find that customers' satisfaction level with the cleanliness of the room (dimension) affects their satisfaction level with the room overall (process). Further, customers' satisfaction level with the room influences their satisfaction level with the hotel overall. Finally, customers' satisfaction level with the hotel affects their willingness to return to the hotel.

Customers' satisfaction levels are assumed to fall into one of three categories: dissatisfied, satisfied, or delighted. Customers' satisfaction levels can move from dissatisfied to satisfied, or from satisfied to delighted. A customer must already be satisfied before it is possible to delight him or her. Because there are two possible shifts in customers' satisfaction levels, the effects of satisfying a customer (shifting from dissatisfied to satisfied) and delighting a customer (shifting from satisfied to delighted) are determined separately.

Using the statistical models for predicting satisfaction and delight, it is then possible to estimate the relative importance of the various processes in predicting satisfaction or delight with the firm, and the relative importance of the dimensions in predicting process satisfaction or delight. These importance weights can then be related to performance, as measured by the percent satisfied or percent delighted. Charting the importance and performance information can help provide preliminary ideas about where resources should be allocated. As you will see later, these approaches also provide the basis for the ROQ model.

PREDICTING REPURCHASE INTENTION

Of all the questions on the questionnaire, the one with the most direct link to customer retention concerns the customer's stated likelihood of repurchasing the product or service. If the information is collected in percentage terms (100 percent, 80 percent, etc.) and if we accept those percentages at face value, then there is an immediate link to market share and profitability. However, it is important to remember that actual customer behavior may differ from the customer's stated intentions. For example, not all of those saying "100 percent" will actually return. But historical data can be used to validate these categories (e.g., 93.4 percent of the "100 percent" category repurchase), restoring an accurate link to market share and profitability. This will be very useful later on in deriving financial implications of quality improvements.

In any event, we are most interested in how overall satisfaction and overall delight with the firm relate to customers' likelihood of repurchasing the service

FIGURE 5-1   EFFECT OF OVERALL SATISFACTION AND DELIGHT

Repurchase Intention

Delighted	95.2%
Merely Satisfied	84.7%
Dissatisfied	31.3%

Effect of Delight   = .952 − .847 = .105
Effect of Satisfaction = .847 − .313 = .534

(repurchase intention). This is easily determined by examining the average repurchase intention of the "Dissatisfied," "Satisfied," and "Delighted" categories. Figure 5-1 shows that Delighteds are likely to return 95.2 percent of the time, Satisfieds are likely to return 84.7 percent of the time, and Dissatisfieds are likely to return only 31.3 percent of the time. From these percentages, you can calculate the incremental repurchase intention impact of moving a customer from satisfied to delighted, or from dissatisfied to satisfied. The incremental effects are obtained by simple subtraction, yielding an added 10.5 percent from delighting a customer, and an added 53.4 percent from satisfying a customer.

PREDICTING OVERALL SATISFACTION

It is nice to know that overall satisfaction has a big impact, but this information is meaningless unless we can figure out what *produces* overall satisfaction. This involves analyzing customer satisfaction data for each business process constituting the firm's service or product.

Separating the effects of satisfaction and delight means creating two new variables, a satisfaction variable and a delight variable, from every three-point scale response. Figure 5-2 shows how to build the new variables. The new variables are constructed as "dummy variables," indicating that their value is either 0 for "No" or 1 for "Yes."

If the response was "1 = Dissatisfied," then the customer was neither satisfied nor delighted. Thus both the satisfaction and delight variables are coded 0. If the response was "2 = Satisfied," then the customer was satisfied but not delighted.

FIGURE 5-2   CONVERTING THE SATISFACTION SCORES TO DUMMY VARIABLES

Raw Score	Satisfaction Dummy	Delight Dummy
1 (Dissatisfied)	0	0
2 (Satisfied)	1	0
3 (Delighted)	1	1

Thus we set the satisfaction variable to 1 and the delight variable to 0. If the response was "3 = Delighted," then the customer was both satisfied and delighted. Both variables are thus coded as 1. (These operations are easily accomplished using the "recode" options of any standard statistical package.)

A data set prepared for analysis will have the general form of Figure 5-3. The respondents (customers) are rows, and the satisfaction and delight variables are columns. The columns should include the overall satisfaction score, the process satisfaction scores, and the process dimension satisfaction scores. They should also include the overall delight, the process delight, and the process dimension delight scores. (The columns do not have to be in the order shown in Figure 5-3, but they must be present somewhere.)

Not every respondent will have answered every question. Therefore, given that some respondent records (or rows) in the data set will be incomplete, a strategy must be devised for dealing with them in the analysis. There are several missing data options typically available in statistical packages. Often the default option (the option that happens automatically unless you override it) is the "listwise deletion" option; that is, any row (respondent) that contains a missing variable is thrown out of the analysis. This is the appropriate missing data option when only one variable is used to predict another variable (bivariate analyses). However, when several variables are used to predict another variable (multivariate analyses), this is often a very bad option, because most of the data set ends up getting thrown out.

Another option that is often used by customer satisfaction researchers for multivariate analyses is "mean substitution," in which a missing value is replaced by the average value for that variable. This option preserves the size of the data set, but at the expense of including some data that do not actually exist. Other, more elaborate methods include "data imputation," which uses the other variables to predict what the missing value would be. This is a sophisticated approach and is not

FIGURE 5-3   PREPARING THE DATA SET FOR ANALYSIS

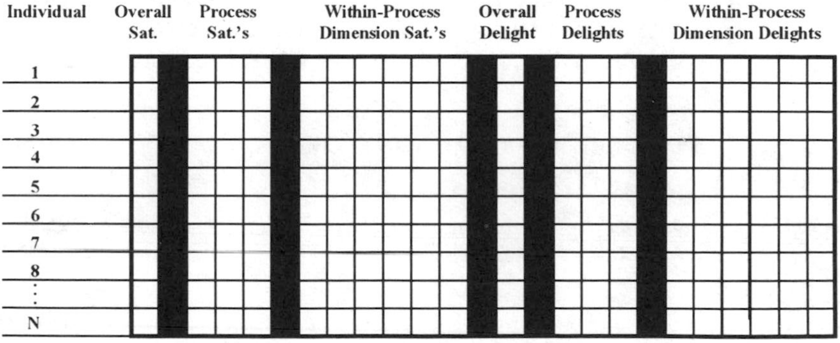

available on all statistical packages. We recommend mean substitution unless a data imputation option is available.

Given that a complete data set has been constructed, there are both simple (bivariate) and more complex (multivariate) ways to relate the process satisfaction levels to the overall satisfaction level. The easiest way is to use a procedure called "regression." Regression is a statistical method that allows you to quantify the relationship between variables. Further, once that relationship has been established, the relationship can be used to predict various outcomes.

Regression can be bivariate (in which case only one independent variable is used to predict another variable) or multivariate (in which case more than one independent variable is used to predict another variable). The predictor variables are called independent variables. The variable being predicted is called the dependent variable, because its value *depends* on the values of the independent variables. Figure 5-4 shows an example of the output that can be expected from running a bivariate regression. You should pay particular attention to three pieces of the output presented in Figure 5-4: the beta coefficient, the R^2, and the t-statistic.

The beta coefficient is the most valuable result to the manager. It tells how much the dependent variable will change for each unit the independent variable increases. In the example in Figure 5-4, the beta coefficient is .5. Therefore, a one-point increase in customers' satisfaction with BILLING results in a 1/2-point increase in OVSAT (overall satisfaction with firm). Since in this example we are actually interested in predicting the shift in customers from dissatisfied to satisfied, the beta coefficient can be interpreted as meaning that 50 percent of the customers that move from dissatisfied to satisfied with BILLING will also shift from dissatisfied to satisfied with the firm overall. Thus, all else being equal, the larger the beta coefficient, the greater the impact on the dependent variable. When multivariate regressions are run, there will be several beta coefficients, one for each independent variable; however, they will be interpreted the same way.

R^2 is a measure of how well the regression model "fits" the data. R^2 can vary between 0 and 1. The closer the number is to 1, the better the regression model is at predicting the dependent variable. Therefore, an R^2 of 0 indicates no predictive capability, while an R^2 of 1 indicates perfect predictive capability.

FIGURE 5-4   PREDICTING OVERALL SATISFACTION—EXAMPLE BIVARIATE RESULTS

Dependent Variable = OVSAT

Independent Variable	Beta Coefficient	Standard Error	t	p
BILLING	.50	.15	3.33	.00

$R^2 = .25$

The *t*-statistic tells you how certain you can be that the beta coefficient is an accurate predictor. The farther the *t*-statistic is from 0, either positively or negatively, the more confident you can be that the beta coefficient is significant. In general, if the absolute value of the *t*-statistic is greater than 2, then the coefficient is considered statistically significant.[2] Also, the *t*-statistic will take the sign of the regression coefficient to which it corresponds, meaning that if the regression coefficient is negative, then the *t*-statistic will be negative and vice versa. In this case the *t*-statistic is positive, which indicates that an increase in billing satisfaction (BILLING) should have a *positive* impact on overall satisfaction (OVSAT).

If the correlations between the different predictor variables are roughly the same, then running separate bivariate regressions with the dependent variable and each of the independent variables individually will do a pretty good job of computing the relative importance of the independent variables (as measured by their R^2 values). However, if two or more of the independent variables are highly correlated with one another, we have a problem known as "multicollinearity." This results in unreliable regression coefficients; therefore we cannot have a lot of faith in the regression model. Unfortunately, severe multicollinearity is a common feature of customer satisfaction data. As a result, using multivariate regression is not a good idea either.

There is an alternative approach that explicitly takes the multicollinearity of the predictor variables into account. This approach uses what is called the "equity estimator"[3] to obtain regression coefficients that control for multicollinearity. A detailed treatment of the equity estimator is beyond the scope of this book. However, the reader is referred to the original source for details. The ROQ decision support system uses the equity estimator to determine its regression coefficients.

An example of output from the equity estimator is given in Figure 5-5. Notice that all predictors are included at once, rather than running one regression per predictor. The interpretation of the equity estimator coefficients is essentially equiva-

FIGURE 5-5   PREDICTING OVERALL SATISFACTION—EXAMPLE MULTIVARIATE RESULTS

Dependent Variable = OVSAT

Independent Variable	Standardized Equity Estimator Coefficient
BILLING	.45
SALES	.50
PRODUCT	.25
REPAIRS	.19

lent to that of beta coefficients. We can see at a glance, though, that SALES and BILLING have the biggest effects, based on the size of the coefficients. Also, the signs are all positive, indicating that all predictors have a positive relationship with overall satisfaction.

PREDICTING OVERALL DELIGHT

The approach here is similar to predicting overall satisfaction, with one key exception. Remember that our psychological framework dictates that it is only possible to delight a customer who is already satisfied. This means that dissatisfied customers are not really prospects for delight, and thus they should be deleted from the analysis. (They should not be permanently removed from the data set—only excluded from this analysis!) Figure 5-6 shows an abbreviated view of the data set, showing which cases should be tossed out. We can see that any respondent who has a 0 for the overall satisfaction dummy variable is by definition dissatisfied and not a candidate for delight. Thus all respondents with 0 on overall satisfaction should be excluded from the analysis. This is easy to do in any standard statistical package.

Figure 5-7 shows an example bivariate regression analysis relating billing process delight (BILLDEL) and overall delight with the firm (OVDEL). In this case the R^2 is very low (.02) and the t-statistic indicates that the predictor variable is not significantly related to the dependent variable. Figure 5-8 shows the multi-

FIGURE 5-6   PREDICTING OVERALL DELIGHT: PREPARING THE DATA

Individual	Overall Delight	Overall Satisfaction	Process 1 Delight	Process 2 Delight	...
1	1	1	1	0	...
2	0	1	0	0	...
3	0	0	0	0	...
4	1	1	0	1	...
5	0	0	0	0	...
.	.	.	.	.	...
.	.	.	.	.	...
.	.	.	.	.	...

(Delete all cases for which Overall Satisfaction = 0)

FIGURE 5-7   PREDICTING OVERALL DELIGHT: EXAMPLE BIVARIATE RESULTS

Dependent Variable = OVDEL

Independent Variable	Beta Coefficient	Standard Error	t	p
BILLDEL	.15	.12	1.20	.20

$R^2 = .02$

FIGURE 5-8   PREDICTING OVERALL DELIGHT: EXAMPLE MULTIVARIATE RESULTS

Dependent Variable = OVDEL

Independent Variable	Standardized Equity Estimate Coefficient
BILLDEL	.18
SALEDEL	.30
PRODDEL	.10
REPDEL	.42

variate version of the analysis, using the equity estimator. We can see that repair process delight (REPDEL) and sales process delight (SALEDEL) have the most impact as predictors.

PREDICTING PROCESS SATISFACTION

You have seen how the process satisfaction scores can be related to the overall satisfaction score. Ultimately, though, you will need to know what drives the overall process satisfaction. Thus you must relate the process dimension scores to the process satisfaction level. Again, you may do this in either a simple (bivariate regression) or sophisticated (equity estimator regression) way.

Figure 5-9 shows the results of a bivariate regression analysis relating the billing accuracy dimension (ACCURATE) to billing satisfaction process (BILLING). The t-statistic indicates that there is a significant relationship, and the sign on the beta coefficient indicates that the variables are positively related. The corresponding multivariate analysis is shown in Figure 5-10. Here we have collected data on only two process dimensions, billing accuracy (ACCURATE) and whether the bill is easy to understand (EZ). Billing accuracy appears to be somewhat more important, based on the relative sizes of the coefficients, but both have a positive impact on the billing process satisfaction level, as we would expect.

FIGURE 5-9   PREDICTING PROCESS SATISFACTION: EXAMPLE BIVARIATE RESULTS

Dependent Variable = BILLING

Independent Variable	Beta Coefficient	Standard Error	t	p
ACCURATE	.32	.09	3.56	.00

R^2 = .10

FIGURE 5-10   PREDICTING PROCESS SATISFACTION: EXAMPLE MULTIVARIATE RESULTS

Dependent Variable = BILLING

Independent Variable	Standardized Equity Estimator Coefficient
ACCURATE	.36
EZ	.24

PREDICTING PROCESS DELIGHT

Again, you must weed out the respondents who are not prospects for delight. Figure 5-11 shows a schematic of the data set, illustrating that all individuals who are dissatisfied with the process overall should be eliminated from this stage of the analysis. As before, you may use bivariate regressions to get a rough idea of the relative impact of the predictors. Figure 5-12 shows an example. In this case it is clear that the accuracy dimension delight (ACCDEL) does not have much impact on the billing process delight (BILLDEL). The t-statistic and the R^2 are very low. The multivariate analysis, shown in Figure 5-13, also shows little impact based on either predictor.

ESTIMATING RELATIVE IMPORTANCE

Calculating relative importance is fairly easy to do, given the results of the bivariate or multivariate regression analyses presented earlier in the chapter. In the bivariate case, assuming the listwise deletion option was used for missing data, Figure 5-14 shows a simple calculation that may be used. An importance score (0 to 100) is calculated by multiplying the bivariate R^2 by the percentage of cases present. In the figure, you see that the R^2 was .40 and 10 percent of the cases were missing. This yields an importance score of 36 ($.40 \times .90$).

FIGURE 5-11   PREDICTING PROCESS DELIGHT: PREPARING THE DATA

Individual	Process Delight	Process Satisfaction	Dimension 1 Delight	Dimension 2 Delight	...
1	1	1	1	0	...
2	0	0	0	1	...
3	0	1	0	0	...
4	1	1	0	1	...
5	0	0	0	0	...
.	.	.	.	.	...
.	.	.	.	.	...
.	.	.	.	.	...

(Delete all cases for which Process Satisfaction = 0)

FIGURE 5-12   PREDICTING PROCESS DELIGHT: EXAMPLE BIVARIATE RESULTS

Dependent Variable = BILLDEL

Independent Variable	Beta Coefficient	Standard Error	t	p
ACCDEL	.01	.13	.08	.45

R^2 = .00

FIGURE 5-13   PREDICTING PROCESS DELIGHT: EXAMPLE MULTIVARIATE RESULTS

Dependent Variable = BILLDEL

Independent Variable	Standardized Equity Estimator Coefficient
ACCDEL	.01
EZDEL	.02

**FIGURE 5-14   CALCULATING RELATIVE IMPORTANCE: SIMPLE
             (BIVARIATE) METHOD**

Bivariate R^2	.40
×	×
% Cases Present	90%
Importance (0–100)	36

**FIGURE 5-15   CALCULATING RELATIVE IMPORTANCE: MULTIVARIATE
             METHOD**

Standard Equity Estimator Coefficient Squared	.25
×	×
100	100
Importance (0–100)	25

This works because the R^2 applies only to the portion of the data that had a valid response (because missing data was not included). Cases are presumably missing because that aspect was not relevant to the respondent. Thus the importance score for *all* of the cases would be the R^2 times the percentage of cases present.

Of course, as discussed previously, if the pattern of multicollinearity is complex, then using the equity estimator is more appropriate. In this case, assuming that the mean substitution option has been used for missing data, then squaring the equity estimator coefficient and multiplying by 100 gives an importance weight on a 0 to 100 scale. An example is shown in Figure 5-15. The resulting importance weights give a good sense of where a manager should focus his or her efforts to get the most impact. Importance weights derived from the equity estimator are an integral part of the ROQ decision support system.

IMPORTANCE-PERFORMANCE MAPPING

Another data presentation device that many managers find appealing is the importance-performance map. The idea is derived from standard quadrant analysis, as has been used for many years in business strategy. In general, this approach argues, we should be most concerned about those issues for which importance is high and performance (typically measured by average satisfaction) is poor. These yield the greatest potential for gain.

SIMILAR TO
PRAEDO...

FIGURE 5-16   IMPORTANCE-PERFORMANCE IN DRIVING SATISFACTION:
OVERALL PROCESS DATA

Process	Importance	% Satisfied
Billing	45	40
Sales	55	70
Product	25	80
Repairs	15	50

We make one key change to this widely used approach. Recognizing that converting satisfieds to delighteds typically requires different programs than converting dissatisfieds to satisfieds, we analyze those things separately. Thus we measure
performance as either percent satisfied, or percent of satisfied who are delighted.
To implement this approach, start with importance and performance data such as
those shown in Figure 5-16, which addresses the issue of converting dissatisfieds to

FIGURE 5-17   IMPORTANCE-PERFORMANCE IN DRIVING SATISFACTION:
QUADRANT MAP

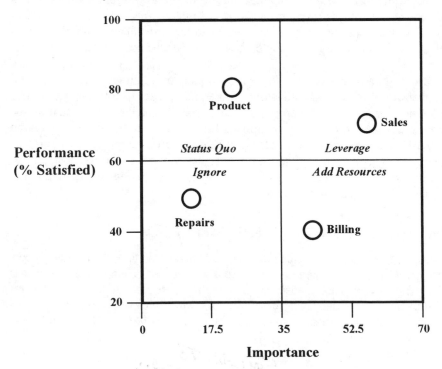

satisfieds; then map the processes in Figure 5-17. The quadrants are somewhat arbitrary but are roughly defined by the averages on the two axes.

The four quadrants can be interpreted in managerially useful ways. The upper-left quadrant is where performance is strong, but the importance is low. At best, this suggests maintaining the status quo. In some cases, there may be opportunities for transferring resources from the processes in this quadrant. The upper-right quadrant is where performance is strong and importance is high. This area represents competitive strengths. We should trumpet our competency in advertising and personal selling. The lower-left quadrant is an area in which we are not doing particularly well, but it doesn't matter. It is best to ignore these areas. The lower-right quadrant represents the area of greatest opportunity. It is important, and we are not doing well. Resources should be added to this area.

One caveat should be mentioned with regard to quadrant maps. Remember that the quadrants are defined in a somewhat arbitrary manner. This means that processes that are near a borderline should be examined carefully. For example, Product is close to the upper-right quadrant, indicating that there may be some leverage opportunities. Also, if there is not much variance on a dimension (e.g., all processes have very high and roughly equal performance), then dividing that dimension may be questionable.

Figure 5-18 shows data for deriving a quadrant map for driving delight (converting satisfieds to delighteds), shown in Figure 5-19. Comparison with Figure 5-17 shows that the satisfaction quadrant map may be very different from the delight quadrant map. Note, for example, that the repairs process changes position on the map. Apparently we are doing something right to delight repair customers, even though we have some trouble getting our customers satisfied in the first place.

The quadrant analysis can also be employed at the dimension level. Figure 5-20 shows the importance-performance map of satisfaction about whether the bills are accurate and easy to understand. It is difficult to draw many conclusions from only two process dimensions in a quadrant map, and, in fact, it is probably a mistake to even draw quadrants with so little data available. Nevertheless, the map shows pretty clearly that making the bills more accurate should be a high priority in improving satisfaction with billing.

FIGURE 5-18   IMPORTANCE-PERFORMANCE IN DRIVING DELIGHT: OVERALL PROCESS DATA

Process	Importance	% Satisfied Who Are Delighted
Billing	18	10
Sales	30	30
Product	10	40
Repairs	14	50

**FIGURE 5-19   IMPORTANCE-PERFORMANCE IN DRIVING DELIGHT:
QUADRANT MAP**

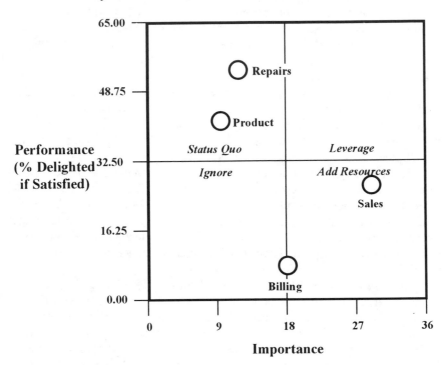

SUMMARY

You have seen that it is possible to derive some very valuable managerial insights from customer satisfaction survey data. The key to deriving useful results is the structure of the questionnaire, which must be organized around business processes. You can also benefit by differentiating between the conversion of customers from dissatisfied to satisfied, and the conversion of satisfied customers to delighted.

Analyses may be conducted in either a simple or sophisticated manner, with greater accuracy naturally being associated with the more sophisticated approach. In either case, importance scores are derived that show the impact of each variable in driving satisfaction (or delight). The importance scores may be mapped in an importance-performance quadrant chart, which yields managerial recommendations about which variables to emphasize.

**FIGURE 5-20   IMPORTANCE-PERFORMANCE IN DRIVING PROCESS
SATISFACTION: QUADRANT MAP**

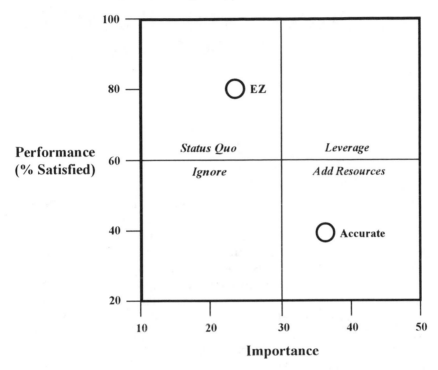

NOTES

1. Abbie Griffin and John R. Hauser (1993), "The Voice of the Customer." *Marketing Science* 12 (Winter), pp. 1-27.
2. The actual number to be used is the "critical value" of t. At the .05 significance level, the value to be used will be slightly less than 2 for large sample sizes and slightly greater than 2 for small sample sizes.
3. Lakshman Krishnamurthi and Arvind Rangaswamy (1987), "The Equity Estimator for Marketing Research." *Marketing Science* 6 (Fall), pp. 336-357.

6

MEASURING THE FINANCIAL IMPACT OF QUALITY

The quality movement has become popular among businesses for one reason: empirical evidence suggests that quality and profits are linked. Unfortunately, while many firms accept that quality and profits go together, few actually track the profits associated with their quality programs. Some managers believe that the value of quality is unknowable, while others do not believe that quality should not be subject to financial criteria. Ultimately, however, whether the profit impact is measured or not, the success or failure of any quality program is its effect on the company's bottom line.

Quality improvement can lead to profits in several ways. The primary sources are: (1) cost reductions from improved efficiency, (2) improved customer retention, and (3) attraction of new customers (see Figure 6-1)

In this chapter we briefly review the overall problem of measuring the financial impact of quality, and describe what practitioners and academics are doing to try to measure the various effects illustrated in Figure 6-1.

THE COST OF QUALITY

Cost reductions from improving processes to do things right the first time were the basis for the initial enthusiasm for quality-improvement processes in the manufacturing sector. The influential work of Deming and Juran focused primarily on process control methods that enabled manufacturers to detect quality problems early in a process, thereby saving the costs of rework and scrap, which can amount to 25 to 30 percent of sales revenue in manufacturing companies and as much as 30 to 50 percent for service firms.[1] Eliminating these wasteful costs can have a dramatic effect on profits.

The resources spent to provide quality on a consistent basis are known collectively as the "cost of quality." Spending usually occurs in four areas:

FIGURE 6-1   MAIN SOURCES OF PROFITS FROM QUALITY IMPROVEMENT

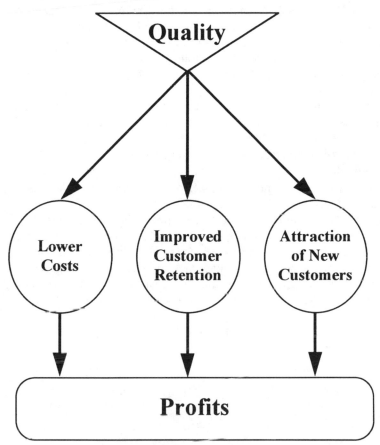

1. Prevention of problems
2. Inspection and appraisal to monitor ongoing quality
3. The cost to redo a defective product before it is delivered to the customer (also known as internal failures)
4. The cost to make good on a defective product after it has been sent to the customer (also known as external failures)

Identifying and estimating these costs can be very difficult, particularly in a service setting. Chapter 7 will discuss the cost of quality in more detail.

THE VALUE OF CUSTOMER RETENTION

The second influence on profits is the retention of current customers. Think of a firm's customer base as water in a bucket, as in Figure 6-2. Customer defections correspond to leaks in the bucket, causing the level of sales to drop. To replace lost customers, the firm must attract new customers, both newcomers to the market and competitors' customers. To make the level rise, the inflow at the top must be greater than the outflow at the bottom. There are two ways to accomplish this: the firm can "turn up the tap" by seeking new customers more aggressively, or it can take steps to plug leaks by keeping its current customers from leaving.

The standard approach has been the former. For the last several decades, businesses have focused on the "offensive" side of marketing through the manipulation of what are called the "Four Ps" to attract and to win new customers. The Four Ps represent the marketing mix variables of Product, Price, Place (distribution) and Promotion (marketing communications). It is implicit in this discussion that the marketing mix must also satisfy current customers, but after-sale service, complaint management, and other key elements of customer-retention programs—"defensive" marketing— have not received as much emphasis.

That is unfortunate. A study by the U.S. Department of Commerce revealed that the cost of winning a new customer is, on average, about five times greater than the cost of retaining a current customer. For example, some magazine publishers claim that subscription price doesn't pay back the costs of signing the subscriber until into the second year. And in mature, increasingly competitive markets, firms are finding that they simply can't assume they will be able to easily find cus-

FIGURE 6-2   MARKET SHARE AS A BUCKET OF WATER

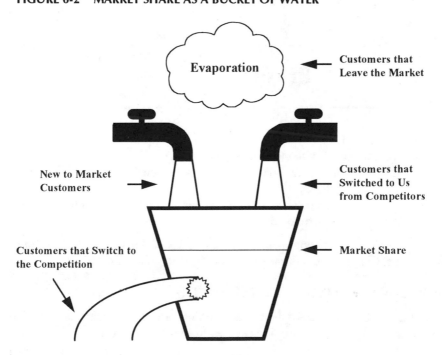

Pull A DIS

tomers to replace those they lose. Customer retention is beginning to receive much more attention. For example, Xerox now includes the opportunity costs of lost sales due to poor quality in its cost of quality calculations.[2]

Bain & Company, a strategic consulting firm, has studied the value of customer retention among many of its clients. Bain claims, for example, that an increase in customer retention rates of five percentage points can increase profits from 25 percent to 80 percent.[3] This is the result of several factors:

1. It's more expensive to win new customers than old ones, as mentioned above.
2. Tracking studies over time have shown that longer-term customers tend to purchase more.
3. In some industries, servicing a familiar customer becomes generally more efficient and, therefore, cheaper.

Bain also finds that there are second-level, though harder to measure, benefits. Long-term, satisfied customers tend to improve the working conditions and satisfaction of the company's employees, so the costs of employee turnover may also be reduced.

High retention is also seen as an excellent competitive weapon. It's not easy for competitors to measure another firm's retention, or even to realize that it is the source of the firm's growth. It's also not easy for them to woo away the firm's highly satisfied customers.

These empirical findings all suggest that investing in quality programs to improve customer retention can be worthwhile, whether it includes programs to improve quality and customer satisfaction initially or programs to address customer complaints. It is no surprise that scientific research has shown that satisfied customers have a higher probability of repurchase than dissatisfied customers over a wide range of product categories.[4] But it has also been found that customers who originally had problems and had them redressed by the firm were almost as willing to return as, and sometimes more loyal than, customers who never had problems.[5]

How Firms Measure the Value of Retention

In spite of the evident value of customer retention, few companies have developed ways to explicitly measure its value. Following are some of the concepts in use.

Problem Impact Tree Analysis

A problem impact tree is a chart that shows the possible outcomes of a customer contact. Either the customer experiences no problem (good) or experiences a problem (bad). If a problem is experienced, then the customer either complains (good) or doesn't complain (bad). If the customer doesn't complain, then there is no opportunity to resolve the problem. If the customer complains, then the problem is either successfully resolved by the company or it is not resolved. Figure 6-3 shows a problem impact tree together with customers' likelihood of repurchasing the service for various outcomes. By using the concept of the lifetime value of a customer together with the probabilities of return at different points of the problem impact

FIGURE 6-3   THE PROBLEM IMPACT TREE

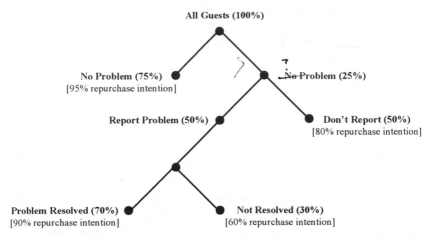

Source: Roland T. Rust, Bala Subramanian, and Mark Wells (1992), "Making Complaints a Management Tool." *Marketing Management* 1 (3), pp. 40–45.

tree, it is possible to determine the expected value of solving or preventing problems. This sort of analysis can give firms a good idea of the scale of solutions that should be undertaken, but it is not particularly diagnostic in pinpointing service attributes that are likely to pay the biggest dividends. It also does not address the value of delighting customers, which is the basis for true loyalty.

Segmentation Schemes

Many industries segment their customers on the basis of demographics or other variables that correlate with lifetime value, retention probability, etc., and use the averages to analyze the amount worth spending on satisfying customers in each group. For example, banks use a number of criteria, such as account age, to segment their customers into groups whose value and retention probabilities are meaningfully different. One banking source says it found a relationship between account value and customer propensity to switch around: the more loyal the customer, the greater the account balance kept in the bank. The average balances of stayers to leavers was 2:1 for accounts less than three years old, 1.5:1 for accounts of four to seven years old, and 1:1 for accounts of seven to twelve years.[6] Loyal customers appear to be worth more than transient ones, even in the short run, and by generally measurable amounts. However, caution should be exercised in imputing cause and effect, since the results don't include information on customer satisfaction with the banks in question. The results could just as well be telling us that customers with large banking relationships find it harder to switch. A complete explanation for the observed relationship probably involves effects in both directions. Other segmentation schemes include identifying differences in retention, repurchase intention, or willingness to recommend across customers who differ in identifiable and manageable ways.

Other Methods of Analysis

Xerox has expanded its cost of quality accounting to keep track of the opportunity costs of lost sales and cancelled contracts, and therefore uses straightforward cost of quality analyses. When the causes of problems are identified, relevant staff attempt to determine the financial value (revenues, costs) of their area if it had no problems and the value of their area with the current level of problems. The difference is the cost of quality for solving this problem.[7] Companies like IBM also now monitor the effect of quality spending on such retention-related measures as installed base and customer satisfaction as part of an overall "quality payback formula."

The debate over the ability to measure the value of quality goes on. Certainly many quality analysts are aware of the value of retention and have developed ways to measure it. However, aside from the types of general, straight-ahead analyses described above, few comprehensive systems for measuring the value of retention have been published in the business press. In Chapter 8, we will describe an integrated program, the ROQ model, which seeks to determine the impact of individual quality programs through their effect on retention rates.

THE VALUE OF NEW CUSTOMERS

The third area in which improved quality improves profits is the attraction of new customers. New customers can become aware of new levels of quality through formal marketing communications or by word of mouth from current customers. Word of mouth is particularly critical for some services, such as banks and other financial services. Research has found that personal referrals are responsible for 20 to 40 percent of bank customers[8] and are a key factor in selecting financial service providers.[9] In fact, many businesses measure "willingness to recommend" as part of their quality tracking program.

For services in which repeat purchases are impossible or very unlikely, such as college degrees and some medical procedures, getting the customer to give an enthusiastic testimonial to other potential customers is the logical equivalent of a retention objective. Organizations that track this measure tend to analyze the extent to which it is correlated with satisfaction with particular attributes to help them direct their quality-improvement efforts toward the most effective programs.[10] However, the financial value of improving the willingness to recommend is very difficult to assess. It is so far removed from the actual appearance at the door of the customer affected that it is probably impossible for the typical firm to determine the expected benefits that are associated with the costs necessary to increase this measure.

Some data are available to indirectly measure the impact and velocity of word of mouth. AT&T Small Business Systems, the division that sells telephone switching equipment to small commercial customers, takes monthly surveys of customer

FIGURE 6-4   TIME LAGS IN THE EFFECT OF QUALITY ON MARKET SHARE

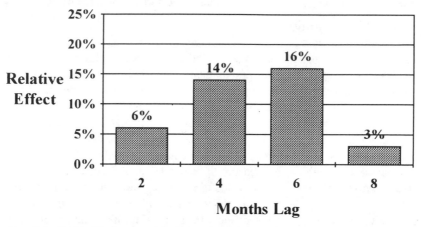

perceptions of the value of its equipment and services relative to those of its competitors. It also tracks its share of new installations by month. Time-series analysis of share and relative quality showed a lagged relationship between quality changes and share, with the greatest increase occurring approximately six months later. Figure 6-4 shows the size of the lagged effects at various intervals.[11] Again, quality appears to generate sales, but the presence of a lagged effect is important. Customers buy systems at all times of the year, and one assumes that AT&T salesmen describe the new changes to potential customers as soon as they happen. So why should there be such a lag? It may reflect the average amount of time for word of mouth, a potentially powerful influence in selling these high-involvement systems, to spread personal testimonials about significant product and service improvements.

However customers learn about quality, there is no shortage of empirical evidence that it has an impact on sales to new customers. For example, the tremendous inroads into the U.S. market achieved by Japanese automakers during the 1970s and 1980s has been attributed by analysts and customers alike to the high quality of their cars compared to American models. Figure 6-5 shows that the same effect has worked for Ford Motor Company. Ford has worked hard to increase the quality of its products, instituting problem-solving work teams in its own plants and encouraging its suppliers to meet rigorous quality standards under its Q1 certification plan. The effort has had a significant success in quality and market share improvements. The graphs show a steady increase in share with a customer-based quality measure, reduction in problems reported by customers (TGW, or Things Gone Wrong per 100 cars) during the first six months of ownership.[12]

FIGURE 6-5   TRENDS IN QUALITY AND MARKET SHARE AT FORD MOTOR CO.

Things Gone Wrong (TGW) per 100 Units
in the First Six Months

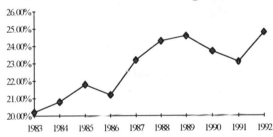

Market Share for Cars and Light Trucks

Source: James A. Welch, Senior Consultant, Arthur D. Little, Inc., Cambridge, MA, July 1993.

CONCLUSIONS

How to measure the likely payback of various possible expenditures on quality is one of the most difficult problems facing managers today. Many firms now realize that tracking financial data alone is insufficient. It generally tells how we have done in the past, but says very little about our likely success in the future. Therefore, efforts are being made to identify the impact on profits of different kinds of quality initiatives. In particular, the effects of quality programs on profits take time to develop, and so some means must be found to predict and track their effectiveness. To anticipate future corporate performance, managers are turning to measures that are seen as leading indicators of financial performance, such as customer satisfaction, competitive quality, customer perceptions of the firm's stature, and market share.[13]

As we have seen, the payback from quality programs can come from many sources and is not always easy to measure. It doesn't help that many firms fail to track easily obtainable data that could greatly assist them in making these estimates. Many firms fail to do exit interviews with departing customers to find out what problems caused them to leave.[14] We have found that even large, sophisticated companies fail to keep sales records that distinguish between new customers, switchers, and repeat customers—records that could be extremely useful in tracking the relative effectiveness of offensive and defensive marketing programs.

However, given the stakes involved, data collection and measurement techniques for certain aspects of the problem are improving. In the next chapter, we will describe some of the basic concepts of measuring the cost of quality. In Chapter 8, we will introduce an interactive computer model that also measures the value of customer retention on profits.

NOTES

1. Lawrence P. Carr (1992), "Applying Cost of Quality to a Service Business." *Sloan Management Review* (Summer), pp. 72-77.
2. Lawrence P. Carr (1992), op. cit.
3. Frederick F. Reichheld (1992), "The Truth of Customer Retention." *Journal of Retail Banking* 13 (4), pp. 21-24; Frederick F. Reichheld and David W. Kenny (1992), "The Hidden Advantages of Customer Retention." *Journal of Retail Banking* 13 (4), pp. 19-23.
4. Eugene W. Anderson and Mary W. Sullivan (1993), "The Antecedents and Consequences of Customer Satisfaction for Firms." *Marketing Science* 12 (Winter).
5. Claes Fornell (1992), "A National Customer Satisfaction Barometer: The Swedish Experience." *Journal of Marketing* 56 (January), pp. 6-21; and Roland T. Rust, Bala Subramanian, and Mark Wells (1992), "Making Complaints a Management Tool." *Marketing Management* 1 (3), pp. 41-45.
6. Peter Carroll (1991), "The Fallacy of Customer Retention." *Journal of Retail Banking* XIII (Winter), pp. 15-20.
7. Lawrence P. Carr (1992), op. cit.
8. Frederick F. Reichheld and David W. Kenny (1990), "The Hidden Advantages of Customer Retention." *Journal of Retail Banking* XII (Winter), pp. 19-23.
9. *Quality as Consumers See It,* prepared for Travelers Insurance Companies by Yankelovich, Skelly and White, Inc., September 1984.
10. For health care examples, see R. Carey and E. Posavac (1982), "Using Patient Information to Identify Areas for Service Improvement." *Health Care Management Review* 18 (Spring), pp. 43-48; and Arch Woodside, L. Frey, and R. Daly (1989), "Linking Service Quality, Customer Satisfaction and Behavioral Intention." *Journal of Health Care Marketing* 9 (December), pp. 5-17. For an analysis applied to graduate MBA programs, see William Boulding, Ajay Kalra, Richard Staelin, and Valarie A. Zeithaml, "A Dynamic Process Model of Service Quality." *Journal of Marketing Research* 30 (February), pp. 7-27.
11. Ray Kordupleski, Roland T. Rust, and Anthony J. Zahorik (1993), "Why Improving Quality Doesn't Improve Quality." *California Management Review* 35 (Spring), pp. 82-95.
12. James A. Welch, Senior Consultant, Arthur D. Little, Inc., Cambridge, MA, July 1993.
13. Robert G. Eccles (1991), "The Performance Measurement Manifesto." *Harvard Business Review* (Jan-Feb.), pp. 131-137.
14. Reichheld and Kenny (1990), op. cit.

7

THE COST OF QUALITY

QUALITY-COST TRADEOFF

One of the ways that quality improvement leads to profits is through cost savings achieved through increased efficiency. Therefore, before a firm can calculate its return on quality (ROQ), it must determine its costs associated with poor quality and what, if any, cost reductions it can expect from its quality-improvement efforts.

The understanding that high levels of quality can lead to cost savings required a shift in management thought. Managers used to believe that, for any process, there was a tradeoff between cost and the resulting product's or service's quality. They believed that it was not economical for any process to reduce defects to zero. Rather, for every process there was a theoretical optimum where the marginal benefit of improved quality was less than its marginal cost (see Figure 7-1).[1]

The problem with the traditional view, however, is that it underestimated or failed to identify all the costs associated with poor quality. In fact, the costs turn out to be much higher than had been believed—by some estimates, 20–30 percent of sales for manufacturing companies and 30–50 percent of sales for service companies.[2] As a result, businesses that have established total quality programs have found that when the full costs of poor quality are recognized, the most economical process is one in which all products produced meet specifications (see Figure 7-2).[3]

Although management has come to accept the fact that higher quality results in lower costs, the value of setting up programs to measure these costs is not universally accepted. While quality gurus such as Juran and Crosby advocate measuring such costs, Deming considers the idea a waste of time.[4] Such programs do provide one important benefit, however: they translate quality problems into dollars. Because these numbers are usually quite large, they can get management's attention

FIGURE 7-1   CLASSIC MODEL OF OPTIMUM QUALITY COSTS

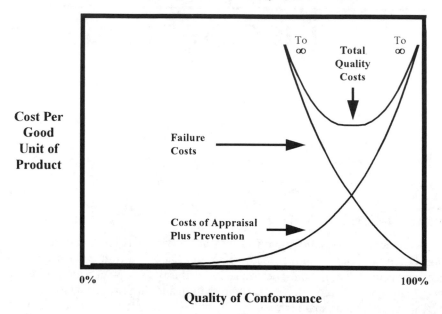

Source: Frank M. Gyrna (1988), "Quality Costs," in Quality Control Handbook, 4th ed. (Juran, ed.), New York: McGraw-Hill. Reproduced by permission of McGraw-Hill.

regarding the need for quality improvement. Further, such programs can cost-justify particular quality-improvement efforts.

COST OF QUALITY

All firms recognize the need to identify and track the costs of their operations. Most firms, however, do not aggressively monitor their quality-related costs. While some quality-related costs are usually tracked, they are seldom fully accounted for, nor are they viewed holistically as the cost of poor quality.

"Cost of quality" programs attempt to determine the financial impact on the company of preventing, testing, or repairing defective products or services.[5] The result is a single dollar figure associated with poor quality. The two main objectives of the program are to quantify the financial consequences of quality problems and to identify areas for quality improvement and cost reduction.

The cost of quality can be divided into four main categories: (1) Internal Failure Costs, (2) External Failure Costs, (3) Appraisal Costs, and (4) Prevention Costs (see Figure 7-3).[6]

FIGURE 7-2   NEW MODEL OF OPTIMUM QUALITY COSTS

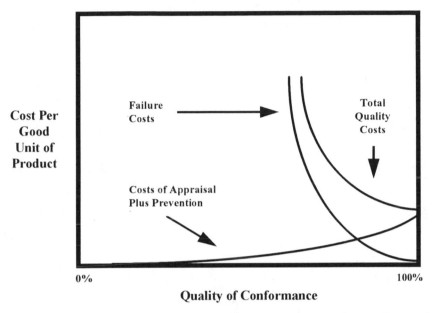

Quality of Conformance

Source: Frank M. Gyrna (1988), "Quality Costs," in Quality Control Handbook, 4th ed. (Juran, ed.), New York: McGraw-Hill. Reproduced by permission of McGraw-Hill.

Internal Failure Costs: The costs associated with errors that are discovered and corrected before the product is delivered to the customer. Examples include scrap, rework, downtime, and discounts resulting from substandard quality.

External Failure Costs: The costs associated with errors that are discovered after the product or service has been delivered to the customer. Examples include warranty and replacement costs, refunds or rebates, and customer complaint handling costs.

Appraisal Costs: The costs associated with ensuring that the product or service meets specifications. Examples include inspections, lab tests, and field tests.

Prevention Costs: The costs associated with avoiding errors. Examples include training, preventive maintenance, and process planning.

All businesses incur costs associated with all four categories of the cost of quality. Although quality expert Philip Crosby has popularized the notion that "quality is free," the fact is that there is no way to reduce the cost of quality to zero. Firms that have not embraced total quality control tend to find that most of

FIGURE 7-3   ASSIGNMENT OF COST ELEMENTS TO QUALITY COST CATEGORIES

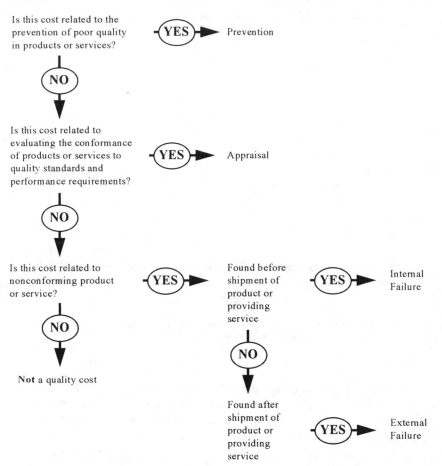

Source: Jack Campanella, ed. (1990), *Principles of Quality Costs,* 2d. ed., Milwaukee: ASQC Quality Press, p. 42.

their quality costs occur in the failure and appraisal categories. Quality-conscious firms have found that it is much more cost-effective to focus on preventing errors.

While focusing on prevention may sound intuitive, that is not the standard procedure in most firms. Usually, when a firm begins experiencing quality problems, its first response is to increase inspection activities. This method fails, however, because the cause of the problem has not been addressed.

One of the reasons companies address quality problems through inspection rather than prevention is that inspection can merely be added to the end of the process. Therefore, inspection requires no real change to a process. Focusing on prevention means redesigning how a company structures its processes. If a company is serious about eliminating quality problems, however, it must design processes that produce error-free products or services. Quality experts claim that 80 percent of defects are caused by design problems or from purchasing supplies based on price instead of quality.[8] This means that the way to get rid of most defects is to prevent them from ever becoming a part of the process.

Eliminating errors makes good economic sense. Firms engaged in total quality control have demonstrated that the cost-benefit tradeoff is in favor of prevention costs over failure costs. In other words, it is more economical to produce a defect-free product than to have to rework and repair the product (see Figure 7-4). The cost savings associated with eliminating defects can be substantial. It has been

FIGURE 7-4   FAILURE COST AS A FUNCTION OF DETECTION POINT IN A PROCESS

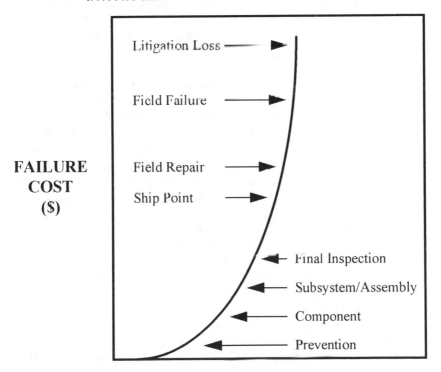

PROCESS

Source: Jack Campanella, ed. (1990), *Principles of Quality Costs*, 2d. ed., Milwaukee: ASQC Quality Press, p. 42.

reported that for most businesses, every dollar invested in prevention could save $10 in internal failure costs and up to $100 in external failure costs.[9]

OBSTACLES TO COST OF QUALITY

Companies interested in measuring their cost of quality frequently find that there are many barriers to be overcome. One of the biggest obstacles is that traditional cost-accounting measures do not fully track or identify costs associated with poor quality. Typically, accounting systems do some tracking of prevention and appraisal costs, as well as internal failure costs. Normally, however, they do not do a good job of monitoring external failure costs. This can mean a substantial under-reporting of the cost of quality, since by some estimates 70 percent of a firm's typical failure costs are due to external failures.[10]

Further, most accounting systems ignore the strategic implications of quality improvement. For example, high defect rates may cause disruption to the production process, longer lead times, and the need to expedite items. Ignoring these strategic costs causes underestimation of the cost of quality. Worse, this can lead to incorrect identification of areas that will provide the maximum benefit to the company from quality improvement.[11]

The way to handle these accounting problems is to use what information is available from the current accounting system. Estimates can then be made for elements not currently monitored. This leads to an important point regarding cost of quality analysis. Cost of quality studies are not designed to replace a company's accounting system. Instead, they are a tool designed to demonstrate the financial importance of quality and to help identify areas where quality improvement can benefit a company financially. Therefore, being reasonably close will serve the company just as well as being exact. In fact, if being precise means using an inordinate amount of time and/or disrupting operations, then approximating these costs will be better for the company.

Another problem companies face is deciding how to allocate uncommon but large costs. For example, a product liability claim may be an unlikely occurrence, but should it happen, the costs would be significant. Including such costs in the cost of quality analysis would distort the figure and may cause management to discount the reasonableness of the entire analysis. Still, it probably should be noted separately as an unlikely but potential cost.

Some categories in a cost of quality study can be controversial. Examples of such categories might include overhead, depreciation on equipment used for inspection, and loss of customer good will. Including controversial costs without agreement from those affected in the company is not a good idea. The cost of quality is usually a large enough figure without including controversial categories.

Finally, firms that are not involved in manufacturing may have difficulty explicitly defining conformance quality. Also, these firms frequently have not had experience using quantifiable performance measures. In such cases it is essential to determine process specifications that affect customer satisfaction. For example, when the U.S. marketing division of Xerox measured its cost of quality, it defined its product as "100 percent customer satisfaction," meaning that anything causing less than complete satisfaction was a defect.[12]

IMPLEMENTING COST OF QUALITY

There are several ways to begin implementation of a cost of quality program. All methods, however, require the commitment of senior management for the effort to succeed. Because senior management acceptance and support is imperative, it is best to begin by demonstrating to them the magnitude of the cost of quality. This requires an analysis of the primary costs associated with providing quality products or services.

This analysis can be conducted through a review of the company's financial data. Where costs are unavailable, estimates can be used. The focus of this review is on uncovering major costs, and does not have to account for all costs associated with poor quality.

The analysis needs to be broken down into two parts: an overview of the entire company and an overview of a particular area of the company. The overview of the company will demonstrate the magnitude of the cost of quality. The overview of a specific operation will show management how to calculate and eliminate these costs. Therefore, when selecting an example area, it is best to select the section that represents the most obvious opportunity for improvement.

After the review has been conducted, the information must be presented to management. The presentation should do the following:

1. Demonstrate the need for action given the size of costs associated with poor quality.
2. Recommend the creation of a task force, chaired by someone in management and including individuals from all major departments, to calculate the cost of quality.
3. Suggest that an area of the organization be used as a pilot study. (The logical area would be the one used as an example in the presentation.)

It is likely that they will be surprised and perhaps skeptical because of the size of the figure. Quality experts have found that it is not uncommon for initial cost estimates to account for greater than 20 percent of sales.[13] Given that the majority of this information will come from their own financial data, however, management will likely accept the findings. This will, one hopes, cause management to agree to create the task force and implement a pilot program.

The first objective of the task force should be to determine the categories that constitute the cost of quality. After a list has been created, management needs to agree to the categories. Getting members of management to agree to the categories used in the analysis increases the likelihood that they will accept the figures presented to them from the study.

Once agreement has been reached regarding the categories, the next step is to assign responsibilities and a schedule for collecting data for the pilot area under study. The data will come from several disparate sources. Examples of such sources include:[14]

Cost accounting data: This includes data on items that reflect quality costs, such as scrap and rework. The main problem with this data is that it could be dated, and therefore may not reflect the current situation.

Payroll data: This includes a full or partial assignment of employees' salaries for work on quality-related activities.

Estimates from knowledgeable personnel: Estimates from appropriate individuals can be sought on such things as calculating the time to repair a product or allocating the percentage of time an employee is engaged in quality-related activities.

After the data has been collected, the information needs to be charted over time. It is wise to collect data for a reasonable period of time before drawing conclusions or planning actions. Because costs can vary with the level of activity, it is important to track these costs in terms of the total dollars used and as some percentage of one or more measurement bases that represent indicators of business activity. Otherwise, it is impossible to determine if cost changes are the result of particular actions taken or simply changes in business activity. Examples of costs of quality measures that reflect business activity include: (1) quality costs per unit produced, (2) quality costs as a percent of sales, (3) quality costs as a percent of cost of goods sold, and (4) quality costs as a percent of total manufacturing costs. It is also important to track quality costs by the category to which they correspond: prevention, appraisal, internal failure, or external failure.

After the quality cost information has been tracked for a sufficient period, the next step is to identify opportunities for improvement. Because prevention costs are less expensive than appraisal or failure costs, the logical areas to focus the company's quality-improvement efforts on are those requiring inordinate appraisal costs or causing internal or external failures.

After the actions have been tracked over time, the results of the program need to be summarized and presented to management. Assuming the results are positive, the presentation should also recommend the following:

- The company's employees should be informed of the results of the program.
- Key members of each department should be educated in the concepts of a cost of quality system.
- The program should be implemented throughout the company.

Getting management to inform the employees of the results of the program should not be difficult, because good news is easy to tell. Further, since the task force contains individuals from all key departments, educating members of the various departments should be relatively easy, since a knowledgeable individual is located in each section. However, although management should commit to implementing a cost of quality program throughout the entire company, it should not move too quickly by installing the program in all areas at once. The program should progress at a reasonable pace, moving from department to department so that the organization can learn from each area and devote attention to the program's success.

One advantage of this approach is that both management and employees can see successful implementation of the program over time and under a variety of circumstances, thereby reinforcing their commitment to the program. Another is that inclusion of all departments in the design and implementation of the program prevents the analysis from becoming viewed as the "product" of the quality department.

LIMITATIONS

Cost of quality analysis is an effective tool for getting management's attention regarding the amount of money a company currently spends to correct or prevent errors. It also shows how shifting spending from appraisal and failure categories to prevention can result in substantial financial benefits. Cost of quality analysis cannot, however, be the primary means of monitoring a company's quality efforts. More timely, nonfinancial measures are needed to establish targets for quality improvement.

There are several problems with relying on cost of quality as a performance measure. First, measuring the cost of quality in and of itself will not solve quality problems. Solving problems requires the action of management. Second, cost of quality programs are vulnerable to short-term management practices. It is often possible to reduce quality costs in the short-term without improving quality. In fact, eliminating prevention and appraisal costs will result in immediate but short-lived cost savings.[15]

Also, the cost figure will almost always underestimate the actual cost of quality. It is difficult, if not impossible, to determine the costs associated with things like customer ill will and plant disruptions. Further, it is impossible to determine an optimum cost of quality level. There will always be some cost associated with providing a quality product or service, but there is no way to accurately determine if too much is being spent. Finally, while prevention costs are generally presumed to be more cost-effective than appraisal and failure costs, there is no optimum distribution of costs over the four categories of cost of quality. As a result, while cost of quality analysis provides important information regarding a company's quality efforts, it is not a good measure of actual performance. Therefore, if an area has been identified as needing improvement from a cost of quality analysis, it is extremely important that nonfinancial methods be instituted to accurately monitor the process.

SUMMARY

Because cost savings are one of the ways that quality leads to profits, companies should determine their cost of quality before calculating their return on quality (ROQ). Quality costs fall into four categories: prevention, appraisal, internal failure, and external failure. External failure costs are the most expensive quality costs to the company, followed by internal failure costs. Prevention costs—quality costs that focus on preventing errors—are the most cost-effective quality costs.

A company's quality costs are usually quite large, often accounting for over 20 percent of a firm's sales. As a result, the size of these costs can focus management's attention on the need to improve quality. Therefore, management should be made aware of the costs the company incurs as a result of poor quality.

When implementing a cost of quality program, it is best to use a task force made up of individuals from all major departments, as well as management. The task force should determine the categories that will comprise the company's quality costs. Once the categories have been resolved, the task force should begin the program by focusing on a particular department to serve as a pilot study.

The pilot program should begin by uncovering the current cost of quality for the area. These costs should be tracked over a reasonable time period before any conclusions are drawn. The results should point to potential opportunities for improvement. The company should then take action on the most promising areas. Once the pilot program has been completed, the results need to be presented to management as well as to the employees. The program should then be instituted companywide; however, this should be done slowly.

While cost of quality is an effective tool for getting management's attention regarding the need for quality improvement and for isolating opportunities for cost savings, it cannot be the primary means of overseeing a company's quality efforts. Therefore, it is extremely important to use nonfinancial methods to monitor these efforts.

NOTES

1. Robert S. Kaplan and Anthony A. Atkinson, *Advanced Management Accounting,* 2d ed. Englewood Cliffs, N.J.: Prentice-Hall, 1989.
2. E. B. Baatz (1992), "What Is Return on Quality, and Why Should You Care?" *Electronic Business* (October), pp. 60-66.
3. Kaplan and Atkinson (1989), op. cit.

4. D. A. Garvin (1987), "Competing on the Eight Dimensions of Quality." *Harvard Business Review* (November-December), pp. 101-109; P. Crosby, *Quality Is Free.* New York: McGraw-Hill, 1979; and W. Edwards Deming, *Out of the Crisis.* Boston: MIT Press, 1986.
5. Lawrence P. Carr (1992), "Applying Cost of Quality to a Service Business." *Sloan Management Review* (Summer), pp. 72-77.
6. Frank M. Gyrna (1988), "Quality Costs," in *Quality Control Handbook,* 4th ed. New York: McGraw-Hill, pp. 4.1-4.30.
7. P. Crosby (1979), op. cit.
8. "The Push for Quality." *Business Week* (June 8, 1987), p. 135.
9. George P. Bohan and Nicholas F. Horney (1991), "Pinpointing the Real Cost of Quality in a Service Company." *National Productivity Review* (Summer), pp. 309-317.
10. Lawrence P. Carr (1992), op. cit.
11. P. Nandakumar, S. M. Datar, and R. Akella (1993), "Models for Measuring and Accounting for Cost of Conformance Quality." *Management Science,* vol. 39, no. 1 (January), pp. 1-16.
12. Lawrence P. Carr (1992), op. cit.
13. Jack Campanella, ed., *Principles of Quality Costs,* 2d ed. Milwaukee, Wis.: ASQC Quality Press, 1990.
14. Frank M. Gyrna (1988), op. cit.
15. Wayne J. Morse, Harold P. Roth, and Kay M. Poston, *Measuring, Planning, and Controlling Quality Costs.* Montvale, N.J.: National Association of Accountants, 1987.

GET CROSBY'S BOOK(S)

8

THE ROQ DECISION SUPPORT SYSTEM

THE MANAGER'S PROBLEM

Quality is not a magic wand for which cost is no object. It is an investment and a tool that must be paid for in the short run for longer-run results. Not every customer-pleasing service enhancement is equally important in building loyalty or in generating cost savings, and consequently, given limited resources to invest in service improvement, spending on quality becomes a classic resource allocation decision.

So how is a manager to decide which areas warrant spending and how much should be spent on them? Many of the data needed to answer these questions are near at hand—in the satisfaction data collected through the questionnaires described in Chapters 4 and 5, in corporate records on the internal costs of poor quality, and in the judgment and intuition that the firm's managers have developed through years of experience. However, pulling all of these facts and judgments together to estimate the consequences of different actions is a daunting task. It would be helpful to have an interactive computer model, known as a decision support system, that pulls all the data together and produces logically consistent estimates and forecasts of the likely financial effects of quality-improvement programs. The model should be designed so that a manager can quickly and easily test different programs and different assumptions about their likely effectiveness.

In this chapter, we describe the necessary components of a Return On Quality (ROQ) decision support system. It incorporates two major sources of profitability from quality programs: cost savings due to internal efficiencies, and higher sales due to higher retention rates of our current customers. (This book contains information on how to order a demonstration version of such a decision support system. It is designed to show how such a system should work. The appendices contain a user's guide for the software, as well as sample cases that demonstrate how such a system can assist managers in making quality-improvement decisions.)

AN OVERVIEW OF THE ROQ DECISION SUPPORT SYSTEM

Much has been written about how improved quality lowers internal costs by reducing rework and increasing other efficiencies. The "cost of quality," as this topic is called, was discussed in Chapter 7. The ROQ model is based on the premise that it is also possible to measure the impact on profits through the improved retention of the firm's current customers due to better quality. There are, in fact, many quality initiatives—such as increased variety at a restaurant—for which improved customer satisfaction, and not lower cost, is the primary effect and primary source of increased profits. Because its goal is to retain current customers, it is called "defensive" marketing. The chain of effects by which quality improvements increase loyalty, repeat purchases, and ultimately profits is shown in Figure 8-1.

While quality improvements can also make the firm more attractive to new customers in the market and to competitors' customers, these "offensive" effects take longer to build, and their timing is far more difficult to predict. They are dependent upon the efficiency of word of mouth among customers and the extent to which the firm spends money on promotion. Therefore, ROQ focuses on the "defensive" impact of quality. This restriction means that ROQ calculations may underestimate the eventual total financial impact of changes due to quality programs. Nevertheless, they provide a conservative lower limit for the value of such programs.

FIGURE 8-1   THE CHAIN OF EFFECTS OF CUSTOMER QUALITY ON PROFITS THROUGH RETENTION

FIGURE 8-2   THE LOGICAL FLOW OF THE ROQ DECISION SUPPORT SYSTEM

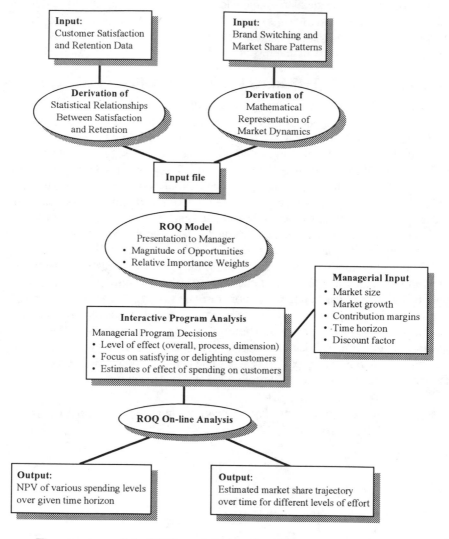

The structure of the ROQ model's inputs and outputs is shown in Figure 8-2. The model is based on the chain of effects in Figure 8-1. The technical details of how the components are linked together are in the appendices.

Linking Customer Satisfaction to Retention

Several of the model's relationships are considered to be outside management's control in the short run, and form the backdrop against which the impact of various quality programs can be tested. In particular, a survey of the firm's customers is required, asking them about how various aspects of the firm's services meet their expectations and their likelihood of remaining a loyal customer during the next period of time. From this information, you can estimate the relationships from (3) to (4) in Figure 8-1 using statistical methods—i.e., how satisfaction and delight drive retention.

Determining the relationship between customers' satisfaction levels and retention is especially important in choosing areas for attention. The most important areas are those that cause customers to leave or to be loyal, not necessarily those that receive the largest number of complaints or those for which satisfaction is rated the worst. You must also estimate the relationships that describe how satisfaction and delight at lower levels of the service drive satisfaction and delight at higher levels. You can then follow the chain of relationships from satisfaction or delight at the dimension level up through retention. The methodology necessary to determine these relationships was described in Chapter 5.

Choosing Between Quality Initiatives

The data should then be analyzed to create a series of importance-performance tables. The tables present the importance of the various dimensions and processes, as well as the percentages of customers who are currently dissatisfied, satisfied, or delighted with each of the dimensions and processes (see Figure 8-3). The importance score indicates how strongly the customer response at any lower level is related to the response at the next higher level (e.g., dimension to process, or process to

FIGURE 8-3   IMPORTANCE-PERFORMANCE TABLE

	Physicians		Nurses		Admitting Process	
	% Delight	Importance	% Delight	Importance	% Delight	Importance
Overall	23%	16	26%	33	6%	12
Dimension 1	14%	23	26%	7	9%	12
Dimension 2	26%	25	15%	22	1%	8
Dimension 3	40%	5	4%	35	6%	22
Dimension 4	5%	19	14%	9		
Dimension 5	12%	1	7%	12		
Dimension 6	18%	13	30%	21		

Note: Importance scores are on a scale from 0 to 100. Importance scores for various dimensions reflect the strength of the relationship between the dimension and the process overall.

firm overall). A firm's quality performance is shown by the percentage of dissatis-fied, satisfied, or delighted customers it has with various dimensions and processes.

Using the method presented in Chapter 5, each importance score is a number between 0 and 100 that indicates relative impact. The higher the score, the greater the impact. By examining information at the firm level, the process level, or the subprocess level, the manager can consider programs that are targeted at narrow aspects of the service or broader programs that affect customer attitudes about processes or even the service overall.

Managers should be careful not to focus only on importance scores. If the firm has a small percentage of dissatisfied customers or a high percentage of delighted customers, the dimension or process with the highest importance score may not represent the best opportunity (because there are few people left to be shifted). Therefore, managers need to consider whether the potential exists to shift a signifi-cant percentage of customers from dissatisfied to satisfied or from satisfied to delighted for a particular dimension or process, as well as the corresponding impor-tance score.

The two sets of information together give the manager some initial indications of where to most effectively allocate resources and the tradeoffs involved. The obvious place to begin would be a critically important service aspect with which many customers are dissatisfied. However, companies with ongoing quality pro-grams will probably have identified such "low-hanging fruit" and be trying to decide among less obvious choices. It is with those choices that ROQ will be most helpful. For example, just because a large number of customers express dissatisfac-tion with some dimension doesn't necessarily make it a priority problem if the dimension has very little to do with customers' satisfaction with the process. On the other hand, a highly important dimension may not be worth further investment if most customers are already delighted with it. Tradeoffs must be made, but ROQ will help in the decision by giving profit information for each alternative.

Linking Retention to Market Share

For a firm to determine the relationship between its customer retention rate and its market share, it must have an understanding of the dynamics of the market under current competitive conditions. Simply tracking revenues, profits, or market share will not provide enough information to link retention to market share. Therefore, companies must track the following information:

- The extent of brand/firm switching by customers in the market
- The rate of entry of new customers into the market
- The percentage of customers that are new to the market that are attracted to your firm
- The percentage of customers who leave the market

These figures are assumed to represent (or must be judgmentally adjusted to represent) the levels of offensive and defensive activity in the market under current competitive conditions. The ebb and flow of customers to and from our brand/firm provide the projections of market share and contributions if no new actions are taken. This will serve as a baseline for comparing the net present value of programs intended to change our retention rate. It provides information for estimating the link from step (4) to step (5) in Figure 8-1.

Figure 8-4 shows how these components fit together to affect market share. Basically, a firm improves its market share by attracting new customers at a rate greater than its market share percentage and/or increasing its retention rate as a result of improving customers' satisfaction levels through improved quality. However, the firm's attractiveness to new customers is assumed to be unrelated to the customer satisfaction of existing customers (which is why the faucets have no valves in Figure 8-4). Therefore, the only way to increase market share through customer satisfaction is by raising the firm's retention rate. Improving customer satisfaction increases the retention rate—i.e., the hole in the bucket in Figure 8-4 gets smaller—which allows the firm to keep a larger percentage of its customers and thereby improve its market share.

FIGURE 8-4   LINKING RETENTION TO MARKET SHARE

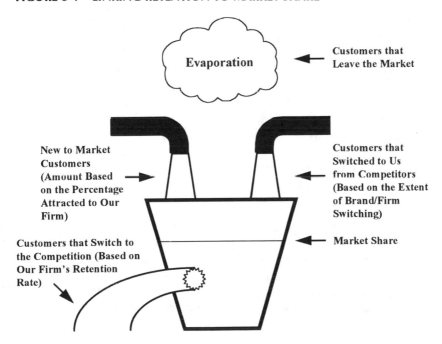

Linking Market Share to Revenues

Once the dynamics of the market are understood, it is relatively easy to link market share and revenues. Two additional pieces of information are necessary: the size of the market and the average profitability per customer of the segment under investigation. By multiplying the market share by the number of people in the market times the average profit per customer, it is possible to convert market share forecasts into dollars—i.e., steps (5) and (6) in Figure 8-1.

Linking Quality Programs to Customer Satisfaction

Once the links between customer satisfaction and revenues are understood, it then becomes possible to estimate how changes in customers' satisfaction levels will affect market share and revenues. What is needed is a way for the manager to explore the profit impact of various proposed quality programs. However, quality programs usually have some cost associated with their implementation (even if ultimately they produce a net cost savings). Therefore, the expected changes in the percentage of dissatisfied, satisfied, or delighted customers must be linked to various levels of spending to improve quality. This information corresponds to the links between steps (1), (2), and (3) in Figure 8-1.

Managers should have some idea what the firm currently spends on various processes and dimensions to "keep customers satisfied" (i.e., reduce dissatisfied customers) and to exceed their expectations (i.e., increase delighted customers). Further, the customer satisfaction surveys show how many customers are dissatisfied, satisfied, or delighted with various aspects of the firm's service. Therefore, the level of customer satisfaction that corresponds to quality spending is known for the current level.

Managers must also estimate the expected impact on customer satisfaction that will result from proposed spending to improve quality. This can be done by using management judgment or by experimenting with the program on a small scale to determine the likely effect.

With this information, the chain of causation in Figure 8-1 is now complete. The firm can now estimate the profit impact of proposed quality programs. It observes the following pattern:

1. The dollar impact of changes in market share are determined for the current competitive situation.

    Because the dynamics of the market have been modeled, projected changes in market share for future time periods (that are expected to occur without additional quality spending) can be translated into dollars. This will serve as the baseline figure from which comparisons are made.

2. The effect of a proposed quality program on customer retention is estimated (based on expected changes in customer satisfaction).
3. The effect of improved retention on market share over time is determined.
4. Projected changes in market share are translated into dollars.
5. Costs associated with implementing the quality program are subtracted.
6. Cost savings resulting from improved "cost of quality" are added to the total.
7. The difference between the baseline profit figure and the new dollar figure is the profit impact of the proposed quality program.
8. To calculate the Return On Quality (ROQ), divide the profit impact by the costs associated with implementing the program.

CONCLUSIONS

The ROQ decision support system attempts to measure what has until now generally been considered unmeasurable—the impact of quality on profits through increased customer retention and market share. The start-up costs to measure this impact are not great, given that a firm is already monitoring its standing among its customers. The program requires data from the sort of customer survey that many firms are already conducting, and other essential measures of competitive performance. In return, the program enables a manager to organize his or her thinking about how quality programs will justify their expense through improved revenues. The program requires some possibly difficult managerial judgments, but these judgments are made anyway, and are implicit in decisions being made without the model. The ROQ model shows managers the logical implications of those assumptions.

Several vivid lessons have been drawn from using the ROQ model on actual corporate data. The first is that quality programs aimed at customer retention take time to pay off. Net present value (NPV) calculations of many programs computed over just one year often show the most profitable course is to spend nothing. However, when the NPV is calculated over horizons of two or three years, spending on retention programs frequently pays off handsomely if given time to develop.

The second lesson is that many programs don't ever pay off through increased retention alone. However, that does not mean they shouldn't be instituted. If they generate suitable cost savings, they may well be worth the expense. Therefore, cost savings must be incorporated into the ROQ calculation.

By using the ROQ model, managers can now rely on more than the belief that quality leads to profits. They can project the expected profit impact resulting from their quality programs. As a result, managers can make sure that the effect of their quality programs shows up on the bottom line.

9

STRATEGIC PLANNING: THE ROQ APPROACH

For over a decade, American firms have tried to emulate the quality programs they saw being developed by their Japanese competitors. Early on, many seized upon the use of quality circles as being the key to salvation, only to be disillusioned when these exercises failed to bring about the hoped-for results, and employees became cynical about being used. Others have sought to adopt statistical process control without adequately training employees to interpret data and find solutions to quality problems. In some cases, quality-improvement efforts have not been adequately translated into customer-defined quality. In all such firms, quality programs inevitably fall into disrepute as yet another useless passing fad.

But even in those firms in which customer satisfaction has become the company watchword, near-religious zeal rather than fiscal responsibility has often driven corporate activities, resulting in a flurry of expensive and poorly coordinated programs that have brought down the wrath of corporate budget keepers. Lavish service may produce happy customers, but the firm's shareholders are often less pleased. In time, as quality managers are unable to justify their activities in tough economic times, these quality programs also become at risk.

What is needed is not just a quality mindset, but a financially accountable quality mindset that focuses on the "return on quality." Moreover, the fiscally responsible management of quality—real customer-defined quality—cannot be achieved by halfhearted, piecemeal, or cosmetic means. Converting a firm to a customer-driven mindset while managing the return on quality-improvement expenditures requires tightly coordinated planning throughout the entire organization, from top to bottom, with the profitable improvement of customer-defined quality as the overall driving force. This objective should define the direction of the organization as a whole and guide the development of specific objectives and strategies at each lower level. In particular, strategies at all levels should be selected on the basis of their contribution to higher-level goals, as well as on their ability to meaningfully and profitably contribute to customer satisfaction.

In the sections that follow, we will illustrate a planning process that is rooted in the concept of return on quality. The sections will review all aspects of planning, but those that are novel to ROQ will be highlighted.[1]

THE PLANNING CYCLE

Whatever the objective, planning should be a continuous process, following the cycle shown in Figure 9-1. Plans are drawn up and implemented; results are then monitored to provide input for the next plan—to either fix problems or to build upon successes. The cycle is actually much more complex than the simple diagram would suggest. There is often a great deal of negotiation between the planning and implementation stages as managers find certain aspects of the plan unworkable and suggest alternative strategies. In fact, while many plans are done on a large scale perhaps once a year, reviews of results are often performed quarterly or even monthly, resulting in possible midcourse corrections to objectives and strategies. This monitoring of programs should be carried out continuously on a smaller scale for programs at all levels of the organization. Shewhart called this the PDCA cycle for Plan—Do—Check—Act.[2]

FIGURE 9-1   THE PLANNING CYCLE

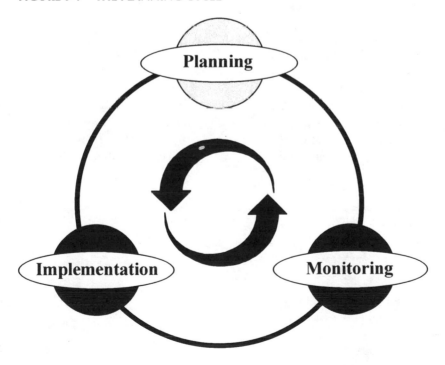

The implication of this cycle is that there are clear, quantifiable objectives to be reached, action programs designed to reach those objectives, and measures to be monitored to determine whether the firm is on track.

But planning for return on quality takes time, given the deep understanding of operations necessary. Figure 9-2 shows the stages that a firm must pass through on its way to becoming skilled at ROQ planning. Initially, managers will have little understanding of the likely effect of quality improvements on customer satisfaction, and the ROQ calculations will depend entirely upon management estimation. How-

FIGURE 9-2   STAGES OF CORPORATE LEARNING

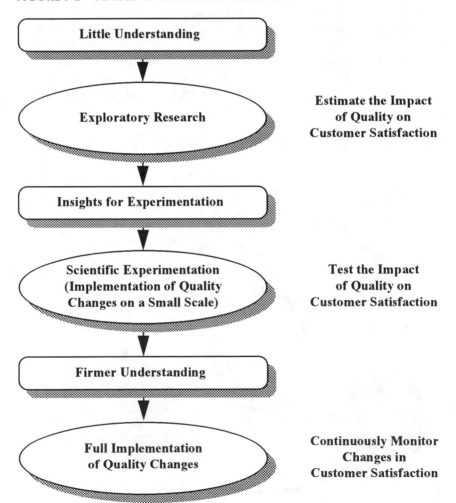

ever, as those estimates—combined with sensitivity analysis—reveal likely areas for improvement, experiments can be run to improve management understanding of customer response. These will lead to more solid estimates, which can be used to guide full-scale roll-outs of programs. The figure simplifies the actual situation, since competitive and market conditions always change and forecasts are almost always wrong. Therefore, past data are never a certain guide to the future, and some managerial estimation is always necessary. Nevertheless, the figure suggests how management must build up a fund of experience over time to improve its understanding of market response to quality improvements.

BENEFITS OF PLANNING

Planning provides several important benefits:

- Planning directed primarily toward achieving a high return on quality continually communicates the organization's need to consider financially justifiable customer satisfaction as the ultimate objective of every employee. Note that this need not be an inhibitor to customer service. Service based on maximizing the lifetime value of customers is usually better than the transaction-focused service with which most of us are familiar.

- Planning also serves as a basis for coordinating activities at lower levels. Employees can see where their efforts fit into the top management's vision for the company and can therefore take initiatives that are consistent with the overall efforts of the firm. ROQ planning ensures that all members of the organization see the service in terms of the processes and dimensions that are important to the customer. It gives them a basis to argue for support and to gain commitment for the programs they see as necessary to achieve organizational goals and objectives.

- The planning structure we will describe forces management to design programs that have an organized, logical basis and are directed toward the firm's overall objectives. In particular, the plan should start with a thorough analysis of the firm's business environment and follow with goals and strategies derived from that analysis to provide the strongest results possible. This logical sequence, depicted in Figure 9-3, is more likely to result in the identification of true sustainable sources of differential advantage and a positive return on quality.

- The plan can serve as the basis for monitoring and control. Those in the organization responsible for achieving particular goals are identified and the data that must be monitored are made clear. The plan forces the firm to think through actions in specific terms, and provides deadlines and benchmarks against which to measure progress.

FIGURE 9-3   THE PLANNING PROCESS

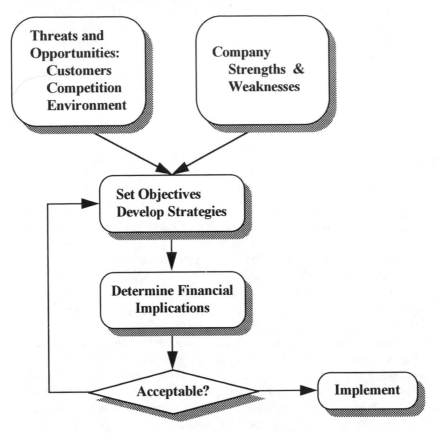

The ideal planning organization, as envisioned by Baldrige consultants Hart and Bogen, does indeed seem like a formidable competitor:

> The strategic value of quality is now recognized at all levels of the organization. Employees know the company's short- and long-term quality initiatives backwards and forwards. Valuable data do not rot in the attic anymore. Instead, information from operations personnel on the shop floor and the sales force in the field is fed back up the organization to become part of the corporate plan. The same thing happens to input from customers and suppliers. No longer internally focused, the company is a true citizen of the marketplace, using world-class benchmarks to drive the goal-setting process. Quality plans confront and root out failure in all functional areas, including support functions. Senior executives are actively involved in mapping out strategy and linking objectives and logistical support. Interlocking quality councils at all levels of the organization speed up the communication process. The company consciously plans for continuous improvement. Episodic and narrowly focused plans are replaced by bold initiatives such as cycle-time

reduction, uniform reduction of defects across the organization, and the improvement of fundamental processes. Long-term quality initiatives, such as meeting educational needs in the work force, are now undertaken.[3]

That does not mean that planning is always successful. Firms that institute planning programs without thoroughly understanding their appropriate implementation and use are likely to generate more problems than they solve.[4] Even Baldrige award winners are capable of falling on hard times, as we have described elsewhere, particularly if the financial consequences of quality programs are not understood in advance. But intuition would suggest that firms with strong planning processes should do better. In fact, although research results are few in number, studies have shown a positive relationship between the level of planning and financial success in organizations.[5] In this context we will describe the process of quality planning and the structure of the planning document.

THE PLANNING PROCESS

The key to making a plan work is less in the contents of the plan than in the manner in which it is developed. The forecasts on which a plan is based are inevitably wrong, meaning that plans always become invalid to some degree. But the exercise of preparing the plan should attune the organization to the requirements for success in its chosen markets and improve its reflexes to respond appropriately to changes. And because delivering quality service requires the involvement of everyone in the organization, it follows that everyone must be involved in some way in the planning process.

Each organization must design a planning process that fits its own culture and personality, but experienced planners and planning consultants agree on several key steps, which will be described next.

Review of Current Plans

Planners must be keenly aware of how well the last plan has fared. Are its objectives being met, and if not, why not? Have customer satisfaction levels with processes and dimensions responded to programs as predicted? Both failures and successes should be carefully noted and learned from to design action programs and research studies for the next period.

Data Collection

The objective of planning is to determine the best course of action possible given the business conditions facing the firm. Therefore, collecting appropriate information about threats and opportunities facing the company is an essential part of any planning exercise. The data needed follow directly from the questions that the plan is intended to answer:

- What are the prospects for profits and growth in this industry?
- What do our customers want?

- Where do we stand vs. our competition?
- What do we have to do to achieve corporate objectives—in particular, profitable customer satisfaction?

Specific areas on which data should be collected are the following.

Market Dynamics

ROQ analysis specifically requires that the firm keep track of, and be able to predict changes in, several key measures. These include the market's growth rate and its components, including the number of new customers and the number of customers leaving the market (the "churn"). External trends that can affect the business should also be tracked and future turning points predicted. These include economic, demographic, and social trends; new advances in technology; and relevant regulatory and political developments. The firm should have ongoing monitoring systems in place to keep track of these important influences in the market and the environment.

Sales Analysis

Gauging the success of quality-improvement programs also requires that the firm be able to determine what part of its growth (or decline) is attributable to various quality efforts. So, in addition to the usual analysis of sales, costs, and profits, managers should also monitor retention rates and brand-switching patterns. All of these data should be available by market segment, by product, by distribution channel, by region—whatever division is likely to provide useful insights to management in identifying problems and opportunities.

Therefore, provision of data should also be a major ongoing part of the firm's MIS function. The importance of this support area is underscored by its being chosen as the second criterion of the Baldrige award. Unfortunately, in many cases this support is not available. Instead, information on products and accounts is collected and organized primarily for cost-accounting purposes and is not readily usable by managers for decision making. We have found major customer-oriented companies that were unable to say what percent of their sales were customer renewals, new customers, or switchers from competitors. A survey of information support systems for marketing managers found the general state of affairs to be "perverse."[6] Therefore, this critical step is likely to be a challenging one for many firms.

External Customers

Since satisfying customers is a major thrust of the plan, managers need to monitor customer satisfaction with current products as well as their future needs and desires, with separate analyses for each key segment. The firm should also be careful that its understanding of each segment's perceptions of the service's constituent processes and dimensions remains current. If market growth and brand switching patterns differ by segments, this should also be noted. Regularly administered ROQ surveys described earlier in the book will provide much of the information, but additional research projects to answer specific questions may also need to be commissioned.

Internal Customers

The firm should also be monitoring employee satisfaction and other measures that provide feedback on the development of empowerment programs and other human resources initiatives. Analysis of employee satisfaction with services provided by other departments should be an ongoing part of quality monitoring, along with more operationally defined measures of support quality, such as number of defects or processing time. Specific unanswered questions may warrant marketing research studies.

Product and Service Quality

Regular tracking of customer reactions to product quality as well as statistical process control of internal quality measures will provide this information, which will indicate areas for particular attention in the next plan.

Suppliers

Measures of supplier quality and assessment of relationships with suppliers will be used to identify quality problems and to determine future quality initiatives.

Competition

All plans must contain thorough analyses of the firm's chief rivals—as defined by customers—to predict what their likely future moves will be. ROQ analysis is particularly concerned with the ability of competitors to attract and retain customers and with the quality initiatives they are taking to raise the general level of expectations among customers in the market in general. Have they matched our quality programs? Have they gone beyond them in ways that will raise the expectations of our own customers or of how the market perceives services to be structured?

Competitive analysis requires a regular program of market structure analysis to identify those firms that customers feel offer the closest substitutes to our products and services. These companies must be monitored to understand their strategies, strengths, weaknesses, and future intentions. Assembling this information requires an ongoing program of competitive intelligence, including activities ranging from the use of clipping services to monitor news about rival firms to primary research to discover their operations, costs, and future plans.[7]

Benchmarks for Processes

To continuously improve its processes, the firm needs to compare itself to the best in the world. Given that the firm is looking for ways to do things better, it must seek world leaders in specific processes, study them thoroughly, and "borrow" ideas that will improve operations. As we have stated before, a firm should not limit itself to its own industry but should study any firm in the world from which it can learn how to improve. The areas to be benchmarked should be closely linked to quality strategy goals so that time and money are not wasted, but there is no limit to the number of areas that may be benchmarked in a quality-driven company. Xerox benchmarks well over 250 different processes, including billing processes at American Express, marketing against Procter & Gamble, and distribution against L.L. Bean.[8]

Environmental Monitoring

Trends in environmental forces must be anticipated, and the likely threats and opportunities they may spawn should be analyzed. These forces include changes in the economy, demographic patterns, cultural and social trends, the regulatory and political climate, and developments in technology. Of particular concern for ROQ analysis is the likely effect of any environmental changes on the size and growth rates of the markets in which the firm competes, on customer needs and the level of their expectations, and on profit margins and the horizon over which the return on investments should be calculated.

Cost of Quality

The costs of poor quality and of preventive measures are essential components of managing the return on quality. Continuous efforts should be made to identify sources of poor quality and to find cost-effective ways of reducing them.

This all amounts to a potentially great amount of data, which could overwhelm the manager who must make some sense of them. In fact, some firms do clog their planning processes with too much, rather than too little, data. Systems for tracking the environment also can be expensive and complex and are not necessarily available to companies with limited resources. Nevertheless, by knowing what sources of information are on the ideal list, a firm that cannot afford them all will at least know what it is giving up and where its decision making may be vulnerable.

Forecasts and Assumptions

As is the case with all planning, ROQ analysis is based upon forecasts, since it attempts to depict the effects of proposed marketing actions. While the analysis of historical data will provide some understanding of likely market dynamics and brand-switching patterns, the future is always subject to change. The purpose of the extensive analysis of environmental trends and competitive behavior is to allow the manager to adjust historical figures to estimate anticipated future changes. Some of these estimates can be obtained from experts at governmental agencies or trade associations, but others may need to be derived specifically for the firm, either internally or by research vendors.

While the complexities of forecasting are beyond the scope of this book, many excellent articles and books on forecasting are available.[9] However, given the high level of uncertainty in environmental forecasts, companies without sophisticated expertise should be extremely careful when using them. To some extent, it is the role of the planning process to supplant the need for precise forecasts by providing a range of contingencies for various possible environmental outcomes.

The step of the ROQ model that requires perhaps the most sensitive forecasts is the estimation of the likely response of customers to specific marketing programs. It is essential that managers be able to provide reasonable values, since their input directly affects the relative financial attractiveness of alternative objectives and strategies. Therefore, during the planning cycle, managers should be encouraged to think in terms of customer responses to marketing activities. Measures that track customer satisfaction with processes and dimensions over time in response to quality-related actions should be provided to help managers develop a feel for the magnitude and time delay of these reactions in their markets. Where possible, experiments should be run to isolate and measure customer reactions to various levels of quality improvement. This deliberate activity is necessary if management is to develop an understanding of how spending to improve quality will affect customer satisfaction and retention rates.

Situation Analysis

Finally, all the data must be analyzed to answer some "so what?" questions. In particular, have our processes slipped over time? How do our measures compare to those of competitors? How are customer needs changing? Are we prepared to meet these new needs better than competitors? What opportunities do we have to significantly and profitably increase customer retention?

The result is often called a SWOT analysis for "strengths, weaknesses, opportunities, and threats." In other words, the firm's abilities must be matched to available opportunities to optimize its prospects for competitive success. The analysis should focus on the key elements that will give the firm a profitable differential advantage with customers, rather than provide a "laundry list" of many unrelated points. The results will serve as the platform from which objectives and strategies are to be developed.

Development of Objectives and Strategies

The heart of the planning process is the setting of objectives and the formulation of strategies based on the data analysis. To guarantee coordination in the many activities planned throughout the organization, it is essential that goals for individual departments and products be consistent with the firm's overall direction and that short-term strategies be positive steps toward its longer-term goals. Therefore, the first step in the planning process is a clear articulation of the company's vision and mission statements and overall strategic objectives, which is made available to all employees. To ensure that return on customer-oriented quality is central to every department's goals, the corporate-level mission and objectives must stress quality, customer satisfaction, and customer retention among the overriding concerns. Once these higher-level, long-term objectives are established, then lower-level and shorter-term objectives can be developed that are logically linked to them.

Planning Roles and Procedures

The planning process must start with the top managers, whose responsibility it is to decide the overall direction of the firm and to establish the values and culture of the organization. However, specific operational details must be the responsibility of those who will have to implement them and who are most familiar with customer demands. Therefore, after the wishes of top management are carefully communicated to the line managers lower in the organization, it is the job of the line managers to set their own objectives and design strategies to accomplish them. Note that if the firm's actions are to be guided by an ROQ perspective, then the dissemination of responsibility for planning requires that decision makers throughout the organization understand the key concepts of customer retention and customer response to quality.

Lots of cooperative vertical communication is necessary throughout the process. Departments and product managers need to let top management know what they're planning, and top management should use its expertise to help lower-level managers set priorities and to guide and suggest improvements before unacceptable initiatives are developed too far. Many firms also encourage the use of cross-functional teams to ensure coordination and cooperation among departments—an excellent practice.

To guarantee that quality is given suitable attention and to provide expert support and coordination among all line and support departments, some companies establish high-level quality councils to which all top executives should belong.[10] Through their highly visible efforts of approving lower-level quality and ROQ goals, reviewing ongoing progress toward those goals, giving public recognition for job well done, and serving on various project teams, executives can contribute greatly to establishing the quality culture throughout the organization.

Objectives

Managing return on quality requires that some objectives specifically address the defensive marketing issues of customer retention and satisfaction with the firm overall and with specific processes and dimensions. In particular, to improve the retention rate of current customers, management must decide whether to emphasize solving problems (decreasing the number of dissatisfied customers) or enhancing services (to delight customers with various aspects of the product or service). These objectives can, in turn, be translated into internal operational objectives once the relationships between operations and customer reactions are understood. All objectives should be quantifiable—i.e., related to specific measures in the plan—and they should have definite time frames.

Determining objectives and strategies is not a linear process. Although some books seem to suggest that objectives are derived from the mission statement in strictly logical fashion, followed by the creative development of strategies, in fact they must be formulated simultaneously. There may simply be no way to accomplish some objectives, so the set of feasible strategies and their estimated financial return must be kept in mind while goals are being set.

From a practical standpoint, care must be taken to choose objectives appropriate to the organization's level of sophistication, particularly if quality principles are new to management. Quality experts often recommend that companies start with objectives that call for small increases in quality and then work up to larger jumps.[11] But don't be afraid to try some big jumps to really stretch the troops. To force truly innovative thinking, objectives that call for performance order of magnitudes better than current levels are necessary. Sometimes bold objectives with no obvious way to reach them can inspire employees to innovate and to discover breakthrough levels of quality delivery that the previously considered set of strategies would never have achieved. For example, a goal of reducing defects by 10 percent is likely to produce suggestions to work harder in the same ways. A goal of reducing defects by 95 percent requires a thorough reanalysis of how processes are performed. But this device must be used selectively. A plan consisting entirely of stretch objectives and unfamiliar strategies is not likely to be taken seriously as a guide to action. Nor are managers likely to be able to supply sound estimates of costs or likely customer responses for the ROQ analysis.

The role of objectives is under debate. Deming insists that they are counterproductive. He assumes that empowered employees will do their best to improve quality anyway, and that pressure to meet specific quantified objectives, often based on very soft forecasts, may only distort their focus away from more productive activities.[12] On the other hand, many managers find objectives a useful way to communicate the firm's priorities and to force employees to think innovatively.

Strategies and Operating Plans

Generating strategies is a mixture of art and science. Strategies should be both appropriate to the objectives they are meant to reach and financially responsible. At the same time, creativity and innovation are essential to break new ground in service design. Strategic planning requires "a combination of trend spotter, number cruncher, and financial planner overlaid with intuitive decisions aided by an emerging kit bag of [Strategic Planning] techniques."[13]

The initial stages of the strategy-generation process should attempt to produce a wide array of possible solutions for consideration. This requires a firm understanding of the SWOT analysis and access to a broad range of ideas. As many sources of insights as possible should be used from inside and outside the firm. Senior management and the quality council should ensure that horizontal communication—i.e., among personnel from different functions and different divisions within the company—takes place to share experiences and ideas. The company should also use its contacts outside the company for fresh perspectives, such as ad agencies, marketing research firms, and consultants who have a good understanding of the company and its markets, but are less likely to be bound by the company line. And, of course, the firm should borrow ideas liberally from the companies it is benchmarking. After managers have studied these sources of ideas, formal brainstorming techniques can be used to generate additional strategies. The temptation to stop as soon as one or two acceptable ideas have been proposed should be resisted. Participants should keep going, pushing harder to generate a wide list of alternatives.

Quality managers have found that the biggest gains can be obtained from projects carried out by interdepartmental teams.[14] Major improvements in quality can rarely be accomplished by single departments. For example, training front-line personnel to be more friendly won't help much if the underlying service is too slow. Spending more on advertising may attract the curious, but won't retain them if the underlying product isn't what they expect. Thus, it is essential that the idea-generation process also be done by multifunctional teams.

From the list of strategies generated, managers must choose those they wish to pursue on the basis of several criteria:[15]

- Is the strategy consistent with the objectives and the SWOT analysis, and does it provide the firm with some sustainable competitive advantage?
- Are the potential rewards of the strategy sufficient to warrant the effort? The ROQ model can assist here in determining the likely financial consequences of alternative programs.
- Are each of the individual programs consistent with overall goals and with those of other departments and corporate functions? For example, plans to increase customer service while reducing personnel may be at cross-purposes unless the system is being fundamentally restructured. Assumptions about costs and returns that underlie the ROQ analysis for one program may be invalidated by the implementation of others. The use of high-level cross-functional teams is intended to reduce this type of problem.
- Is the program feasible given the company's strengths and weaknesses and those of the competition? Do the personnel who must carry it out understand it and believe in it?
- How solid are the key assumptions on which the strategy is based? Does the strategy's success depend upon uncertain outcomes, such as major changes in economic trends or exceptional performance by departments of the company? If so, how acceptable are the negative consequences if the assumptions are wrong? Can the strategy be reversed?

Other Parts of the Plan

The remaining parts of the plan include financial projections and contingency plans. The financial section should include *pro forma* income statements and cash flow projections for the firm, given the proposed strategies. Output from the ROQ model can be used to develop the revenue and profit projections for proposed defensive strategies.

Contingency plans should be drawn from the discussions that selected the strategies to be implemented. The purpose of this section is to have strategies ready to go should key planning assumptions fail to materialize. A great deal of time will be saved later if planners sketch out alternative actions while everyone is deeply engaged in the planning process and institutional understanding of the SWOT analysis is at its most intense.

Review and Approval

The process of review and approval should be an ongoing one throughout the planning process. Most firms still require a formal presentation to and approval by corporate leadership, but in the ideal process this session is a review and fleshing out of already agreed-upon strategies, rather than an initial sales presentation to senior management. The final presentation should be largely ceremonial with no surprises. Developing plans in the lower echelons without the guidance and participation of senior management is wrong for the same reasons that designing products without input from engineering, manufacturing, and marketing is wrong. Both are done in the "real world," but they are extremely wasteful, because the process can go too far before critical errors are identified, causing time-wasting rework. Continuous communication between upper and lower echelons is equivalent to small-batch processing in manufacturing. Problems are detected and fixed early, resulting in a timely, fully acceptable final product. Upper management and the quality council should be kept informed and provide guidance throughout the development of the plans, and drafts should be reviewed by other parts of the organization.

Implementation

Planning is one thing; actually making the plan work is another. A survey of executives at Fortune 100 companies found that a majority of them felt disappointment and frustration with their planning systems, mostly due to problems in implementation of their plans.[16] Much of this book has been concerned with steps firms can take to manage their return on quality.

Monitoring

A well-crafted plan should have clearly defined areas of responsibility and deadlines, so that sources of shortfalls or unexpectedly effective performance can be identified. However, spotting trouble or locating exceptional success requires continuous monitoring. Measurements to be tracked should be clearly specified in the plan itself, since in many cases they provide the operational definitions of what is expected of employees. Some data are available from standard secondary sources, such as sales reports and other basic accounting documents. Some are available from outside vendors, such as Nielsen store audits and other market-monitoring firms. In general, these measurements are related to the objectives stated in the plan.

In addition to tracking the usual barometers of the progress of the business plan, such as sales, profits, market share, response to advertising, etc., quality programs obviously require that internal quality measures be charted and analyzed to spot difficulties and opportunities for improvement. For tracking the return on quality, the following regular monitoring programs of external customers are also recommended:

- Satisfaction measures: overall, with each process and with each of their dimensions, particularly in those areas where estimated customer responses indicated changes in levels of dissatisfied or delighted customers.

- Switching patterns, including rates of customer retention and percent of new customers, and other "background" values of the ROQ model.

In addition, the firm should be monitoring:

- Employee satisfaction with levels of empowerment, compensation and rewards, working conditions, etc., tracked through anonymous surveys.
- Internal customer satisfaction, using surveys that measure the level of support services provided by departments for each customer.

AN OUTLINE OF A PLAN

Every organization has its own ideas about the specific format its planning documents should follow, depending upon what issues need particular emphasis in its industry, the sophistication of planning in the organization, and how the planning responsibility is delegated. In this section we will describe the general contents of a planning document that embodies the principles of customer orientation and return on quality espoused throughout this book. The specific order and format of the plan's structure is not important and should be shaped to the needs of the individual user, but the basic issues addressed by the plan are essential to be sure that profitable, customer-oriented service is maintained.

Plans should be read and used. The specific length of a plan depends upon a firm's situation, but a document of under 20 pages is probably too sketchy, whereas a document of over 50 pages may be too intimidating to be "user-friendly."

The writing style of a plan should maximize communication between writer and reader. In particular, prose is not necessarily the best way to communicate complex technical information. Use outlines and bullet points to organize your ideas. Break up the text with lots of subheadings to make it easy for the reader to locate specific pieces of information. Present numerical data in tables and graphs rather than in long expository paragraphs. Think of it this way: If you were in a hurry to catch a bus, would you want to receive a beautifully written treatise on departure and arrival times, or would you rather have a bus schedule, which typically contains no prose at all?

The typical strategic business plan is organized according to the following sections.[17]

Executive Summary

This section is a concise summary of the plan's major recommendations and their anticipated results. It should be considered the primary document submitted to senior management. The rest of the plan provides support and operational details for the executive summary, should anyone need to consult them, but reading this section alone should provide a clear picture of what the plan says. The summary for a document of this size should be no more than three or four pages in length.

Historical Update

This section should bring the reader up to date on important trends that have determined the current state of the industry and offer some insights about where it might be heading. Possible areas to cover include significant trends in:

- The market's size, growth, profitability, costs, etc.
- Competitive activity, including changes in quality-related activities, pricing approaches, distribution channels, promotion methods, etc.
- Changes in the nature of demand and supply
- Influences of technology
- Changing political and regulatory climate.

Situation Analysis

This section presents an analysis of pertinent data to understand the current and future dynamics of the market. This analysis serves as a platform to support the choice of objectives and strategies. The list of topics to consider includes:

1. Industry Attractiveness Analysis
   - How big is the market, its growth rate, stage of the product life cycle, the average profit level and its variance over time?
   - What competitive forces put pressures on profit levels in the industry, including the costs of quality programs and the prices customers will pay for quality? Extend the concepts of noted strategy expert Michael Porter to include return on quality concerns:[18] the intensity of interfirm rivalry, the threat of entry by other firms, the amount of negotiation leverage possessed by suppliers and buyers, threats of substitute products, and the amount of slack capacity in the industry.
   - What is the future of relevant trends in the environment, such as demographic shifts, technological development, and political and regulatory restrictions, and how will they affect the firm's ability to profitably improve customer satisfaction?

2. Customer Analysis

   This section presents an analysis of the factors that attract and retain customers to provide the support for customer-oriented actions. First should come a clear description of the major segments into which the company separates its customers:
   - Who are they?
   - How do they use the product?
   - What motivates them to buy?
   - What is their buying process?

- How often do they buy?
- What is the average contribution margin per period of customers in this segment?

An analysis of the effect of marketing on each segment should follow.

- What are the switching patterns in this segment? Over the planning horizon, what are the expected growth and mortality rates of the segment?
- What offensive initiatives are necessary to attract new customers from the segment?
- An ROQ analysis should also be done for each segment to determine current satisfaction levels and to reveal which service processes and dimensions have the greatest effect on customer retention.

3. Sales Analysis

   What percent of profit increases have been due to market size growth, market share growth, price increases, cost reductions, and productivity improvements?[19] How do profits and costs differ by product, segment, distribution channel, territory, etc.?

4. Competitor Analysis

   This section presents an analysis of key competitors. It should summarize their objectives and current strategies; their strengths and weaknesses on the dimensions that will affect their ability to compete effectively; and their likely future actions. This section is often presented in tabular form, with short, concise summary entries describing each dimension of the included companies.

5. Self-Assessment

   To make the competitive analysis meaningful, a similar analysis of the company should be done. Brutally honest candor is necessary here, although this can be politically sensitive. You must have an objective side-by-side comparison of your firm and its competitors if you are to choose realistic goals and effective strategies.

6. Forecasts and Assumptions

   Any other forecasts of important environmental and market factors that are necessary to proceed with ROQ strategy development should be included in this section as well.

Objectives

Objectives must be consistent if they are to lead to success. Therefore, the plan should present the entire "cascade" of objectives, from those at the corporate level to those of the various functional areas. Therefore, the plan should list:

- The firm's mission statement/vision
- Corporate objectives

- Objectives for lower-level units, if the plan is being done for a division or subunit of the corporation
- Functional area objectives (e.g., management, marketing, operations, finance, etc.)

Strategies and Program Details

This section describes the action programs that have been chosen to reach the various objectives. Strategies should be broken down into their component parts. In particular, marketing, management, operations, and finance should be covered in this section.

The marketing strategy section should describe how the firm will attract and retain customers. Strategies should be formulated by market segment, if possible, to keep customer service foremost. However, concern for production and marketing efficiency may also require that production and service delivery be coordinated across segments. The profitability of selected programs should be analyzed using ROQ analyses.

The management strategy section focuses on how senior officers will lead the company. Strategies should include leadership- and management-improvement plans. Special care should also be given to employee recruitment and training programs.

Operational strategies describe how operations will be managed, as well as what materials and equipment will be used. This section should discuss quality measurement programs, and programs to measure its cost.

Financial strategies detail how money needs will be met with programs to raise debt or equity.

Financial Statements

This section projects the financial consequences of the proposed strategies under the assumptions about the future presented earlier. It should include budgets for all proposed marketing, human resources, and operations programs and pro forma statements that include projected costs, revenues, and profits. Supporting tables should include cost of quality analyses and ROQ analyses of the likely profit impact of marketing programs.

Monitors and Controls

This section should discuss the variables that need monitoring to guarantee service quality and to manage its financial return. Examples of such measures include (1) measures of marketplace performance, (2) data necessary to improve the use of ROQ analysis, (3) customer satisfaction measures, and (4) internal quality measures.

Contingency Plans

This section should provide a quick sketch of the basic strategies to be employed in the event that key assumptions of the primary plan should prove to be in error.

SUMMARY

This chapter is intended to emphasize those aspects of planning that are most important for developing a strong customer-oriented culture that is also financially accountable. In so doing, we have given only sketches of the contents of certain sections, such as the financial projections, while, on the other hand, providing suggestions for objectives and measures and controls that may be more extensive than a given organization may choose to use. The chapter is not intended to be a self-contained planning template, but rather a complement to books devoted to planning, which tend not to emphasize return on quality.

The success of ROQ-focused strategic quality planning depends upon how well the process is deployed throughout the organization—a very difficult task in most companies. Successful deployment requires new thinking about the role of employees and how they are rewarded. It requires teaching new skills to managers. And it requires a new cooperative spirit among employees so that they are willing to put the greater good of the firm ahead of individual goals.

This culture is evidently not a natural one for many American workers, given the problems that many firms have had in sustaining TQM (total quality management) programs. This communal spirit is at odds with two common characterizations of American society. On the one hand, Americans are often described as independent and self-absorbed, ready to jump to a new company for personal advancement. On the other hand, companies have long operated on principles that prescribe rigid hierarchical management structures and unbending work rules, a culture that fosters distrust and confrontation on both sides.

Leading the revolution to overcome these barriers requires the sincere, tireless commitment of the CEO in all statements and actions. He or she should continuously endorse strategic quality planning; and, if not actually play a role in designing the system, then certainly take part in regular reviews of the process; and provide funding, time, and other support necessary to ensure high-quality planning activities. However, even a committed top manager cannot impose an ideal planning structure on an organization. All planning interventions have political ramifications.[20] Some employees will lose power and others will gain. Long-standing power relationships will be affected. Therefore, when designing the planning process, the personalities and management styles of the firm's personnel and the firm's current stage of development on the long road toward a textbook planning culture must be kept in mind.

It takes time to instill a sound planning process into the culture of an organization. Doing it well requires total commitment from the entire organization and lots of communication, both vertically and horizontally. It requires commitments of time and resources. And it takes patience to learn how to do it properly, because it is unlikely to be satisfactory the first few times. A specialist in writing business plans for small firms claimed that even for such limited exercises it takes at least three years to gain proficiency at writing plans.[21] For a large organization requiring cooperation from many participants, the learning process will certainly take time.

NOTES

1. For an excellent treatment of marketing planning in general, see Donald R. Lehmann and Russell S. Winer, *Analysis for Marketing Planning*. Homewood, Ill.: BPI-Irwin, 1988.
2. W.A. Shewhart, *Economic Control of Quality Manufacturing Product*. New York: D. Van Nostrand Company, Inc., 1931.
3. Christopher W.L. Hart and Christopher E. Bogan (1992), *The Baldrige*. New York: McGraw-Hill, Inc., p. 122.
4. Malcolm H.B. McDonald (1990), "The Barriers to Marketing Planning." *The Journal of Services Marketing* 4 (Spring), pp. 5-18.
5. Stanley F. Stasch and Patricia Lanktree (1980), "Can Your Marketing Planning Procedures Be Improved?" *Journal of Marketing* 44 (Summer), 79-90; McDonald, 1990, p. 7, op. cit.
6. Thomas Bonoma (1984), "Making Your Marketing Strategy Work." *Harvard Business Review* 62 (March-April), pp. 69-76.
7. See Leonard Fuld, *Competitor Intelligence: How to Get It—How to Use It*. New York: John Wiley & Sons, 1985.
8. Hart and Bogan (1992), op. cit.
9. E.g., John C. Chambers, Stinder K. Mullick, and Donald H. Smith (1974), *An Executive's Guide to Forecasting*. New York: John Wiley & Sons; David M. Georgoff and Robert G. Murdick (1986), "Manager's Guide to Forecasting." *Harvard Business Review* 64 (January-February), pp. 110-120.
10. G. Howland Blackiston (1988), "A Renaissance in Quality." *Executive Excellence* (September), pp. 9-10.
11. Blackiston (1988), op. cit.; Hart and Bogan (1992), op. cit.
12. W. Edwards Deming, *Out of the Crisis*. Cambridge, Mass.: MIT, 1986.
13. Ross and Siverblatt (1987), op. cit.
14. Blackiston (1988), op. cit.
15. E.g., see George S. Day, *Strategic Market Planning: The Pursuit of Competitive Advantage*. St. Paul, Minn.: West Publishing, 1984.
16. Joel E. Ross and Ronnie Silverblatt (1987), "Developing the Strategic Plan." *Industrial Marketing* 16, pp. 103-108.
17. For other detailed descriptions of planning documents, see Donald R. Lehmann and Russell S. Winer (1988), *Analysis for Marketing Planning*. Homewood, Ill.: BPI/Irwin; and Karsten G. Hellebust and Joseph C. Krallinger (1989), *Strategic Planning Workbook*. New York: John Wiley & Sons.
18. Michael E. Porter, *Competitive Strategy: Techniques for Analyzing Industries and Competitors*. New York: The Free Press, 1980.
19. Malcolm H.B. McDonald (1990), "Ten Barriers to Marketing Planning." *The Journal of Services Marketing* 4 (Spring), pp. 5-18.
20. McDonald (1990), op. cit.
21. Charles J. Bodenstad (1989), "Directional Signals." *Inc.* (March), pp. 139-141.

10

CREATING A
HIGH-PERFORMANCE
BUSINESS

This chapter was co-written by N. Laddie Cook, James A. Welch, and P. Ranganath Nayak of Arthur D. Little, Inc.[1] in conjunction with the authors.

ROQ analysis shows how customer satisfaction can be used as a driver for improving overall company profitability. Further, ROQ allows managers to determine which customer-focused processes will have the greatest impact on customer retention. This in and of itself, however, will be of little benefit if the implementation of a firm's overall improvement programs is ineffective. What is needed is a framework for applying ROQ analysis. This chapter describes a heuristic framework for implementing ROQ.

Many companies around the world have tried to achieve high performance in their business over the last decade by embracing a range of "solutions," including mergers and acquisitions, downsizing, and total quality management (TQM). Overall, the results have been disappointing. While some of these approaches have proven useful in certain instances, few have led to sustained high performance.

To understand why, consider this analogy: Imagine that a company's business is long-distance running. A doctor consulted to improve running performance says, "The way to win the marathon is by strengthening the left leg." A second doctor says, "No, you must have the right mindset if you want to win." A third doctor advises lung exercises. In fact, many things must work well at the same time. And then the firm must continue improving them—and the way they interact—forever. In short, there is no "magic bullet," no single focus area, that can ever provide sustained high performance for a firm. The formula for success is a balanced approach aimed at satisfying multiple, sometimes conflicting objectives simultaneously.

THE HIGH-PERFORMANCE BUSINESS MODEL

The High-Performance Business model, developed by Arthur D. Little, Inc., appears in Figure 10-1. Helping shape objectives and priorities are three primary groups of stakeholders whose satisfaction is absolutely critical: customers, employees, and owners. Clearly, the organization must also satisfy other stakeholders, such as the communities around it, various governments and regulators, suppliers, and the environment. But unless the firm has met the needs of the first three groups, it will not be able to meet the needs of the others. Product and service quality is a key *customer* stakeholder need, while earning and improving financial returns is a key *owner* stakeholder need. *Employee* stakeholders require equitable compensation, benefits, and opportunities for career growth.

The core of the model consists of the work processes of the company. Getting these processes right is the primary way to satisfy the stakeholders. This is consistent with ROQ's focus on processes and dimensions (subprocesses). Focusing on

FIGURE 10-1   THE HIGH-PERFORMANCE BUSINESS

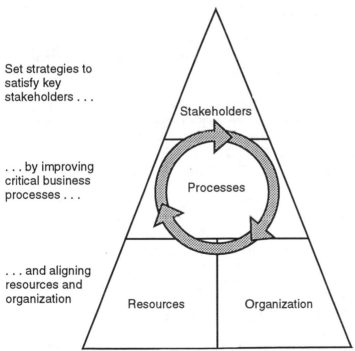

Set strategies to
satisfy key
stakeholders . . .

Stakeholders

. . . by improving
critical business
processes . . .

Processes

. . . and aligning
resources and
organization

Resources   Organization

Reprinted with permission from the first quarter 1992 issue of *Prism*, the quarterly management journal published by Arthur D. Little, Inc.

processes runs counter to traditional thinking, which looks at functions and depart-ments and tries to figure out how to improve them in isolation. The processes of work matter more than the organizational structure; and so, to improve stakeholder satisfaction, the firm has to improve processes. This chapter focuses primarily on how to go about doing that.

At the bottom of the model—the foundation level—are resources and organi-zation. For processes to work well, the firm must have the right resources in place to execute those processes. Similarly, it is critically important to analyze organiza-tional characteristics, which provide the contextual motivation for people to per-form the processes well and to improve those processes at a faster rate than the competition.

Traditionally, the concept of strategy emphasized broad market needs (critical success factors or bases of competition). A more compelling definition involves choosing how to satisfy all key stakeholders—and how to continuously improve their satisfaction.

SATISFYING CUSTOMERS, EMPLOYEES, AND OWNERS

How Stakeholder Satisfaction Works

As ROQ emphasizes, customer satisfaction can translate into profitability and growth. Satisfied customers come back. Further, they tend to buy more and ascend the product line to more profitable products and services. As a result, selling costs decline as well.

To be successful, a firm must deliver consistently high customer satisfaction while simultaneously satisfying its owners and employees. Figure 10-2 shows how satisfied customers lead to fewer customer defections, which lead to high profits and growth, which lead to owner satisfaction. Satisfied owners are more likely to invest in the human resources of the firm—not necessarily by paying employees more, but by providing the training, the equipment, and the conditions to make the work more productive and enjoyable. That leads to a dedicated work force, which, in turn, leads to superior products and services, and that leads back to higher cus-tomer satisfaction. In short, satisfying customers, employees, and owners is a never-ending chain that reinforces itself—a "virtuous circle."

It is important to distinguish between employee satisfaction and employee empowerment. Increasingly, we find that employees value other aspects of the work—such as having a promising career path, professional training, and a compa-ny they can be proud of—ahead of employee involvement or empowerment.

While satisfying key stakeholders sounds fairly straightforward, the fact is that most companies do not attempt to do it in a sustained manner. The case of Wallace

FIGURE 10-2   CUSTOMER, OWNER, EMPLOYEE SATISFACTION

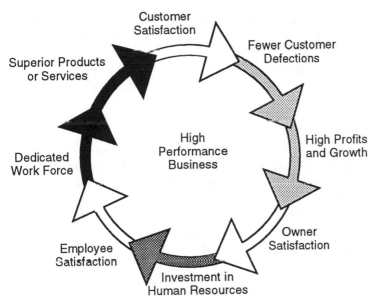

Reprinted with permission from the first quarter 1992 issue of *Prism,* the quarterly management journal published by Arthur D. Little, Inc.

Company was presented earlier in the book. This firm excelled in customer satisfaction, as evidenced by its earning the Malcolm Baldrige National Quality Award. However, although Wallace satisfied its customer stakeholders, it was unable to link customer satisfaction to the bottom line; therefore, it did not satisfy its owner stakeholders, and entered Chapter 11 bankruptcy only months after being presented with the award.

Another example is Florida Power & Light, also mentioned earlier in this book. In the late 1980s, FPL had perhaps the most pervasive program of self-directed work teams of any U.S. company, as measured by the number of team participants and the number of active teams. While this program may have satisfied the employee stakeholders, owner stakeholders were not being satisfied. Since the period of mass empowerment, two separate, sweeping layoffs have been waged by management to reduce costs.

IMPROVING PROCESSES

Consistent with ROQ, the High-Performance Business model focuses on business processes, because these are the vehicles by which key stakeholders are satisfied. A business process is a sequence of connected activities that take place inside, and sometimes outside, an organization, along which information or hardware flows, with something added at each step. The end point of any process is something intended to satisfy one or more of the stakeholders.

Why Focus on Processes?

Viewing an organization as a collection of processes rather than as a rigid departmental structure is particularly useful, because to improve satisfaction it is necessary to optimize the organization as a whole, and processes are the relevant, logical framework for analyzing a firm's effectiveness across the entire organization. The product development process, for example, cuts across research, engineering, manufacturing, and marketing (see Figure 10-3). In traditional organizations, problems

FIGURE 10-3   THE PRODUCT DEVELOPMENT PROCESS

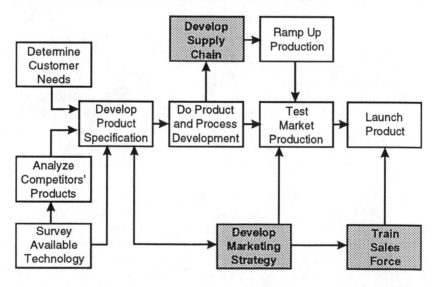

Reprinted with permission from the first quarter 1992 issue of *Prism,* the quarterly management journal published by Arthur D. Little, Inc.

are identified with individual departments, whereas the actual root cause of the problem is often both broader and deeper in the organization.

All business organizations (even not-for-profit organizations) have a number of critical processes that are closely correlated with stakeholder satisfaction. The scope of such processes is often so large that it may cut across both functions and hierarchical levels within the organization. The strategy development process, for example, involves all levels in a company, and the supply chain management process cuts across a variety of organizations: the firm's suppliers, their suppliers, the company itself, and its customers.

These large, critical processes tend to evolve incrementally as companies grow and mature, until nobody understands how the whole process works. And when something is not understood as a whole, it is impossible to improve it systematically. Instead, only little pieces of it can be adjusted or fixed. Because these large and critical processes have become so unwieldy—and are so rarely managed—they vary widely in performance. Figure 10-4 compares Honda's product development process to that of one of the Big Three U.S. automakers in terms of various performance measures. Honda executes this process in two-thirds the time required by its U.S. rivals and with one-third the effort in terms of engineering man-hours. Furthermore, Honda is significantly better in terms of customer satisfaction as measured by the customer satisfaction index.[2] Performance differences of this magnitude are quite typical between the best and the average in many industries, typifying the contrast between high-performance companies and mediocre ones.

Naturally, some processes are more critical than others in terms of their impact on stakeholder satisfaction. We use a systematic analysis to determine which processes within the firm connect to which stakeholder satisfaction attributes.

Different Approaches to Fixing Processes

Once the firm has identified its high-priority processes, it needs to determine which of at least two alternative approaches to improving them is most appropriate: incremental improvement or reengineering (see Figure 10-5).

Incremental improvement consists of small, often sequential improvements in current work processes, typically as a response to problems identified by self-empowered work teams. Incremental improvement can yield results quickly, but these results are usually insufficient to make a substantive difference. There is evidence, however, that the cumulative effect of continuous improvement—if pursued relentlessly and systematically for many years—can be very powerful.

Reengineering a process, in contrast, can result in dramatic improvement in a much shorter time than continuous improvement. Indeed, reengineering can produce results that continuous improvement never would. In essence, reengineering forces management to examine its work processes and ask, "Do we really need this process at all?" For example, many firms have outsourced processes as seemingly

**FIGURE 10-4   AUTOMOTIVE PRODUCT DEVELOPMENT:
HONDA VS. A U.S. MANUFACTURER**

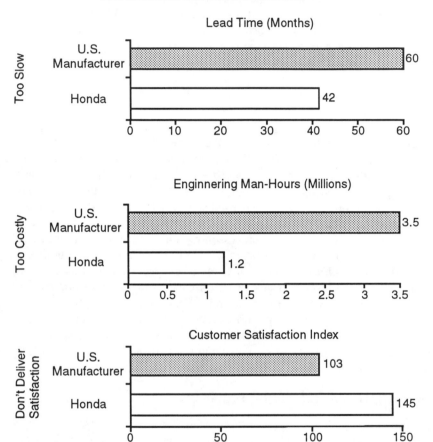

Reprinted with permission from the first quarter 1992 issue of *Prism*, the quarterly management journal published by Arthur D. Little, Inc.

peripheral as mail room operations or as apparently central as the manufacturing process itself. If the firm determines that the process is one it should execute itself, management must examine it as a whole and introspect, "Can we do this better?" Reengineering must be done by midlevel managers who collectively own the entire process. Reengineering allows a firm to re-conceive its entire way of doing business. Reengineering, rather than continuous improvement, is in order when stakeholder satisfaction is extremely poor or when there are conflicts management cannot resolve within its current way of running the business.

FIGURE 10-5   APPROACHES TO FIXING PROCESSES

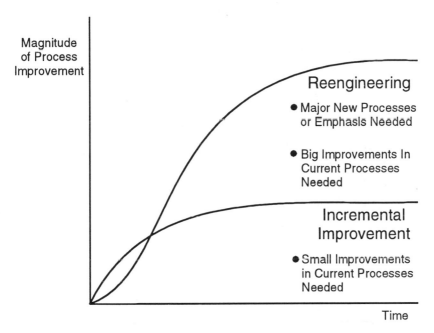

Reprinted with permission from the first quarter 1992 issue of *Prism,* the quarterly management journal published by Arthur D. Little, Inc.

For example, in the banking industry in the early 1980s, it was evident that customers were dissatisfied with banks' working hours and service. In response, banks took very different approaches. Most large banks undertook comprehensive operations improvement studies and opened more branch offices, kept their branches open longer, added tellers, and installed electronic drop-off desks. Customer satisfaction went up, but so did costs.

BayBank of New England took another approach. It probed a little deeper into customer requirements and came to the conclusion that what customers wanted most was access not to the bank but to their money. BayBank was one of the first regional banks to deploy automated teller machines (ATMs) widely. BayBank raised customer satisfaction significantly—at much lower cost than competitors. BayBank's approach to this industrywide problem was truly effective reengineering, while the approach of the other banks was not.

Another example of reengineering comes from the familiar area of just-in-time manufacturing. Traditionally, conventional wisdom held that service quality, as measured by the cycle time to fill a customer's order, was fundamentally in conflict

FIGURE 10-6   REENGINEERING CAN MOVE THE ENTIRE CURVE

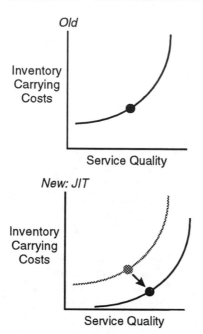

Reprinted with permission from the first quarter 1992 issue of *Prism,* the quarterly management journal published by Arthur D. Little, Inc.

with cost considerations, such as inventory carrying costs. This is a classic stakeholder conflict: The owner's requirement is lower inventory carrying costs, while the customer's requirement is better service.

Business experts spent three decades developing operations research methods aimed at locating the "best" point on this indifference, or "tradeoff" curve. Meanwhile, in Japan, companies developed and refined the just-in-time approach, which substitutes better planning and execution for inventory. This reengineering thinking shifted the tradeoff curve to a different part of the map by allowing businesses to simultaneously improve service quality and reduce inventory carrying costs (see Figure 10-6). Whereas small changes in these tradeoff curves can yield some improvements, reengineering can achieve gigantic leaps forward.

The Perceived Conflict Between Reengineering and Continuous Improvement

Reengineering and continuous improvement can and should coexist in any company. They are not mutually exclusive; rather, they are complementary. Any process that is reengineered will require continuous improvement in the steady-state mode. Furthermore, since reengineering can consume critical resources in the transition phase, there is a definite limit to how many reengineering projects any company can absorb concurrently. For these reasons, continuous improvement will continue to be necessary and important. The key is to choose the best approach for the situation.

ALIGNING RESOURCES

To execute any process well, a firm needs the right kinds and levels of resources in place. Resources are composed of five categories: (1) people, (2) facilities, (3) information, (4) technology, and (5) suppliers. Figure 10-7 lists, for example, resource needs for the product development process. A recent MIT study of management practices among automobile manufacturers in Japan, the United States,

FIGURE 10-7   SELECTED RESOURCE NEEDS FOR THE PRODUCT DEVELOPMENT PROCESS

1. People	Computer-Literate Engineers
2. Facilities	Equipment to Speed Manufacturing of Prototypes
3. Information	Up-to-Date Intelligence on Customers' Needs and Competitors' Plans
4. Technology	Techniques to Analyze Reliability
5. Suppliers	Suppliers with their own World-Class Product Development Processes

Reprinted with permission from the first quarter 1992 issue of *Prism*, the quarterly management journal published by Arthur D. Little, Inc.

and Europe identified, among other conclusions, the absolutely critical importance of having these resources in balance.[3] It is illustrative to compare General Motors' automobile assembly plant in Hamtramck, Michigan, with Toyota's NUMMI plant in Fremont, California.[4] The Hamtramck plant relied heavily on automation based on the theory that it was best to get people out of the plant, because of the belief that people were the source of problems. When the plant opened, the automation went haywire: Robots spray-painted each other, and automated guided vehicles ran into cars. Because of these glitches, production lagged capacity for several months. The company eventually concluded that it had to put skilled, motivated people back into the plant to program, run, and maintain the machines. What resulted was a medium-quality production facility with high costs, because they ultimately needed the labor that GM had "automated out of the process." Furthermore, the capital investment was irreversibly committed.

Toyota approached its design of NUMMI much more thoughtfully. Management recognized that there are things that people do well and others that machines do well, and they blended them together effectively. The plant has less automation, less capital investment, and more highly trained people than Hamtramck. The result is one of the lowest-cost, highest-quality automobile assembly plants in the United States.

ALIGNING THE ORGANIZATION

As the firm fixes its processes to meet stakeholders' needs and aligns resources with the processes, it also needs to look at three aspects of the organization: structure, policies, and culture.

The historical approach to organizational structure has been to group together individuals with like skills doing like work. This is similar to the functional layout of a manufacturer's machine tools where all the drill presses are situated together, all the lathes are together, and the milling machines sit side-by-side. The emphasis was on efficiency, economies of scale, and ease of resource allocation. However, this method is inflexible and focuses on tasks rather than processes. World-class manufacturers have found that to obtain needed flexibility and to produce "economies of scope" (economical production with variety), machines need to be arranged with a product focus so that all the machines required to make one product are arranged in process sequence. As the process focus begins to develop in any business process, the internal boundaries of the old structure need to become porous, so that people sharing a process can work well together. Planning, budgeting, decision-making, and reporting systems all need modification.

In terms of policies, four types really make a difference: (1) performance measurements, (2) salary and reward systems, (3) job scope, and (4) training. These policies have the biggest impact on staff motivation to make critical processes work well and to continuously improve them. Unfortunately, these policies are too often complex and, therefore, difficult to modify appropriately.

For example, the importance of teamwork is recognized in most businesses. Yet very few Western companies have performance measurement and reward systems that are based on teamwork. In Japan, in contrast, this subject has received considerable attention.

Finally, cultural impediments to change must be identified and overcome. We have a simple definition of culture: "the *unwritten* rules of the game." Policies, in contrast, embody the *written* rules. In the culture of many organizations, honest communication is difficult, teamwork is ineffective, and the messenger who conveys the bad news is shot. The roots of such behavior lie in the unwritten rules, which can and must be modified.

When the organization has improved its processes and aligned its resources, structure, policies, and culture, it achieves something that can be called "acceleration," referring to the heightened pace of continuous improvement. For example, while U.S. and European cars have improved over the last 30 years, Japanese cars have improved much faster.

The point is that continuous improvement is not enough: What matters is how *fast* the firm is improving. And the only way to achieve acceleration is to align structure, policies, and culture to dramatically increase people's motivation to make this happen.

GETTING STARTED

To create a High-Performance Business, a firm must satisfy its key stakeholders by improving critical processes and aligning resources and organization to support those processes (see Figure 10-8). This is a major undertaking. Furthermore, it is not something that can be done once and then forgotten. It must be an objective that is pursued forever.

Most companies find radical change difficult, mainly because they do not take the time early on to understand likely resistance to change and where it will come from. People resist change for perfectly understandable reasons, including fear of losing employment, fear of losing power, fear of uncertainty, conflicting messages from management, and doubts about management's long-term commitment to the new way of doing things.

It is essential to unearth those kinds of resistance and to deal with people's concerns in an honest and forthright manner. One way to get support rather than resistance is to involve those who will be most affected in developing and executing the program for change. First, there must be management commitment and leadership, because people look for that. If management does not explicitly show them that this change process is owned at the top, people will be skeptical about it.

Management then must get the involvement of employees. This is best done with facts, because facts overcome emotions. The more decisions about new

FIGURE 10-8   CREATING A HIGH-PERFORMANCE BUSINESS

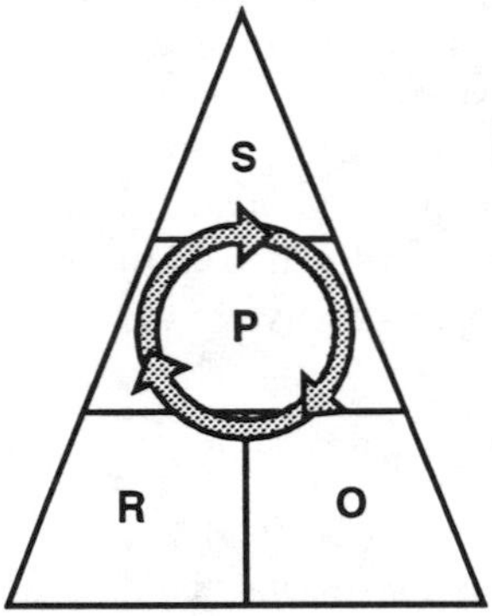

S

Set Strategic Objectives by:
- Listening Actively to all Stakeholders
- Measuring their Satisfaction
- Resolving or Balancing Conflicts Among
  Stakeholders' Needs

P

Focus Process Improvement Efforts by:
- Selecting Business Processes that are
  Critical to Strategy
- Setting Targets for Improvements
- Implementing Incremental, Redesign, or
  Rethink Approaches to Improvement as
  Appropriate

R

Align Resources by:
- Using Technology Aggressively to Achieve
  Superlative Processes
- Balancing Technology and Human
  Resources
- Deploying Flexible Resources to Facilitate
  Continuous Improvement

O

Align the Organization by:
- Modifying Structure to Support Key
  Processes
- Establishing Policies to Motivate People to
  Accelerate Improvement
- Overcoming Cultural Barriers to Integration
  and Change

Reprinted with permission from the first quarter 1992 issue of *Prism,* the quarterly management journal published by Arthur D. Little, Inc.

approaches are based on facts, the more impartial they become, and the change process can go forward with much less friction.

To establish the credibility of the entire improvement process, it is helpful to do an early pilot project. A pilot project demonstrates actual results.

What About Training?

Large-scale change requires training key people. Training needs to be "just-in-time and task-aligned." Just-in-time training means that training in techniques such as problem solving should be provided when the team is about to start an improvement project. Task-aligned training means training specific to the task at hand. These principles are critical. Companies have wasted countless millions of dollars providing training just because it seemed the right thing to do. But when companies train people and then do not give them the opportunity to apply the training within the next few months, people forget what they have learned and their behavior does not change.

It is important to do some very thorough planning before actually starting the implementation process. First, top managers must agree on the vision of transforming the company into a High-Performance Business and understand that it is necessary and desirable.

The next step is the assessment process, culminating in the identification of the first process that will be addressed. Choose a process for which both the opportunity for improvement and the probability of success are reasonably high. If the pilot project fails, it will kill the whole program. Once the pilot project is completed, begin to develop a comprehensive implementation plan.

CONCLUSION

The High-Performance Business model has been used to shape firms' improvement strategies, enabling them to move onto increasingly higher plateaus of performance. Any company can achieve acceleration. And regardless of where the company is now, by internalizing the High-Performance Business model it can overtake the best in its industry.

NOTES

1. N. Laddie Cook is a senior consultant in the Automotive Practice at Arthur D. Little, Inc. Ms. Cook has been assisting her clients—international vehicle manufacturers and their suppliers—to address planning, strategy, and market issues for the past 19 years.

   James A. Welch is a senior consultant at Arthur D. Little, Inc. He has worked with a number of clients in various industries to lead them in the hands-on, day-to-day execution of reengineering critical business processes, primarily in the areas of procurement, operations, and manufacturing. He earned his MBA from Vanderbilt University's Owen Graduate School of Management.

   P. Ranganath Nayak is a senior vice president of Arthur D. Little, Inc., responsible for the firm's worldwide consulting practice in operations management. He has extensive experience helping firms around the world improve their operations—particularly in the areas of research, development, and manufacturing.

2. The customer satisfaction index is an automotive industry metric used to measure customer satisfaction compiled by J.D. Power & Associates.

3. James Womack, Daniel Jones, and Daniel Roos, *The Machine That Changed the World.* New York: Rawson Associates, 1990.

4. NUMMI is the acronym for New United Motor Manufacturing, Inc., a joint venture between Toyota and GM. NUMMI produces the Toyota Corolla and the Geo Prizm.

11

CONCLUSION

The goal of this book is to make quality improvement programs an integral component of corporate strategy. For this to be successful, however, firms must be able to maximize the return from their investments in quality. As a result, our focus has been to show firms how to calculate their return on quality. ROQ analysis is the result of a systematic process. The process begins with the organization viewing itself as a service designed to fulfill customers' needs and engaging in exploratory research to determine those needs. These needs are then related to the company's internal processes to forge a link between company actions and customer needs.

Next, the firm conducts quantitative customer research. This requires collecting data on customers' satisfaction with the various business processes and relating those data to customer retention.

Management must then determine the shift in customer satisfaction with the firm or a business process resulting from a particular quality-improvement initiative. A decision must be made as to whether quality-improvement efforts will focus on shifting dissatisfied customers to satisfied, or satisfied customers to delighted. After the firm has determined the shift in customer satisfaction it expects from its quality-improvement efforts, it can estimate the customer retention rate that would be associated with that level of satisfaction. The organization can then project the market share impact corresponding to the new retention rate. Finally, the company calculates its return on quality by determining the net present value of the revenue generated from the shift in market share plus the cost saving associated with the quality-improvement effort minus the cost of the quality program.

The failure of many acclaimed quality-improvement programs demonstrates that while quality may be necessary to be profitable, it is not a guarantee of profits. Therefore, if managers are going to embrace quality improvement as a critical component of their firms' strategic objectives, then the financial implications of quality must be demonstrated. ROQ allows firms to evaluate their quality-improvement alternatives based on their profit impact. As a result, companies can manage their scarce resources and direct their spending where it counts most.

THE ROQ DEMONSTRATION SOFTWARE

USING THE DEMONSTRATION SOFTWARE

The best way to understand how the ROQ decision support system works is to use it. For this purpose, this book contains information on how to order a demonstration version of such a decision support system designed solely to accompany these appendices. The software provides a tangible representation of how an ROQ system should work. We will refer to the software in this section, and we recommend that the reader run the model on a computer while reading this section. The ROQ software is easy to use and is menu driven. Queries about information always appear in dialog boxes either in the middle or at the top of the screen. Function keys that help you to move through the program and to start various procedures are defined at the bottom of the screen. Extensive help screens are available at every stage of the program, if desired. Details of some of the technical aspects of the program are described later in this appendix.

Before running the model, all files on the accompanying diskette should be copied to the hard drive of your computer in the directory of your choice. Also check the software users' guide located in the appendix for information on system configurations necessary to run the program. When ready to run the program, enter the directory where you have copied it and type ROQ. A startup screen will appear while the program loads. The file is rather large, and takes awhile to load.

Monitor Type

You must next tell the program what type of monitor you have. Select the appropriate type from the menu using either the up and down arrow keys, or by typing the underlined letter in your choice, and then hit [Return].

Relationship vs. Transaction Models

The next menu asks you to choose between the Transaction version and the Relationship version of the ROQ model. For example, the data set we will be analyzing is for a hotel chain, Cumberland Hotels. A hotel's interactions with its customers are discrete events. Customers must make a brand decision every time they go to a hotel, and each *transaction* poses the risk of losing the customer through poor service. (Therefore, select Transaction.) The measure of retention asked on customer surveys refers to the likelihood of the customer choosing a Cumberland Hotel the next time he or she must stay in a hotel in a city where Cumberland is one of the choices.

On the other hand, for many services such as banks, long-distance telephone companies, and insurance agents, the relationship between firm and customer is more of a long-term partnership. Individual interactions will generally not result in brand switching, even though the customer may find some of them dissatisfying. The complex nature of the relationship and the difficulty in switching one's account to a competitor may make customers more tolerant in the short run. However, even with the difficulties involved, there may come a point when the customer has had enough bad service and switches. The ROQ model can accommodate this type of relationship by setting the number of transactions per period to 1, and by asking the retention question on the customer's survey as "What is the likelihood that you will still keep your business with this firm next year?"

ROQ uses different assumptions about the arrival rate of new customers for the two different models, and must therefore know which use you intend.

Choosing a Data Set

Two sample data sets have been included with the ROQ model. For now, choose the data set called "Hotel" from the menu. These data are from the Cumberland Hotels Case, which will be discussed later in more detail. The structure of the service is portrayed in Figure A1-1. At this point we will only use the data to illustrate the workings of the model.

Expert vs. Novice

You must also indicate whether you are a "Novice" or an "Expert." If you claim to be a novice, the standard help screen for each stage of the program will appear without prompting each time you first enter that stage.

THE REVIEW SCREENS

Overall Satisfaction and Delight

The first data review screen presents some tabulations of customers' overall ratings of the hotel. We see that 13 percent of the surveyed customers were classified as

FIGURE A1-1   THE SERVICE STRUCTURE OF CUMBERLAND HOTELS

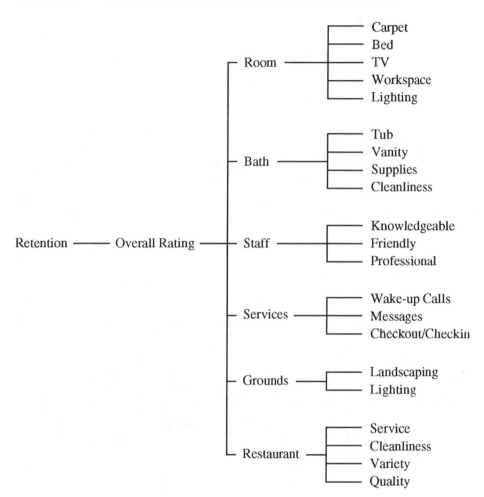

DISSATISFIED with the hotel overall (i.e., found the service "worse than expect-ed"), 65 percent were SATISFIED ("about as expected"), and 22 percent were DELIGHTED ("much better than expected"). In particular, the program suggests that there are two courses of action possible to improve customer retention: One can fix problems to convert the dissatisfied customers to satisfied ones, or one can institute programs that could delight the currently satisfied customers. The ROQ model assumes that these tasks are generally very different, and therefore treats them as separate activities.

Dissatisfaction with Units and Dimensions

Press the function key F2 to advance to a review screen that describes the percent dissatisfied with the hotel service's first five units and their various dimensions.

You can go back by pressing F3. Try it now, and return by pressing F2. The units described on this screen are Room, Bath, Staff, Service, and Grounds. They are further broken down into five, four, three, three, and two constituent dimensions, respectively, as shown in Figure A1-1.

For each unit and for each of its dimensions, two values are given. The first is the percent of customers answering the question who claimed to be dissatisfied with that aspect of the service. For example, 17 percent of those who rated their room overall found it worse than expected. As for the various dimensions of the room, 10 percent were dissatisfied with Dimension 1 (the carpet), 9 percent with Dimension 2 (the bed), etc. The information on dimensions will presumably give us some clue about how to improve the overall rating for the room, just as breaking the service into separate units will help us to improve overall ratings of the hotel.

The second number given for each unit and dimension is an importance score. These values range from zero to 100 in value and are measures of the extent to which the item in question is statistically related to the same measure at the next higher level. For example, the importance score of 38 for Room Dimension 3, the TV, indicates a moderate statistical relationship between the satisfaction with the TV and overall satisfaction with the room. The importance score of 31 for Room Overall indicates the strength of the relationship between satisfaction with the room and satisfaction with the hotel overall. If we assume that these relationships are causal, the importance score of 38 suggests that we can modestly influence customers' overall satisfaction with their rooms, and ultimately their overall satisfaction with the hotel, by improving satisfaction with the TV sets. On the other hand, Room Dimension 1, Carpets, has an importance score of 0, indicating that there was no positive statistical relationship between satisfaction with carpets and satisfaction with the room.

Press F2 again to advance to a screen showing information only about the five units. The ROQ model is able to accommodate up to 20 dimensions per unit, and this screen is available to describe any dimensions beyond the 10th. Because none of the units in our example has over 10 dimensions, this screen is essentially empty. Pressing F2 once more presents the basic satisfaction information on the sixth and last of the units in the example, the Restaurant and its four dimensions. Pressing F2 again shows the empty screen available to accommodate possible higher dimensions of Restaurant.

Delight with Units and Dimensions

Advancing once more with F2 brings us to a screen describing delighted responses to the first five units and their dimensions. The percentages and importance weights are similar to those for satisfaction. For example, 17 percent of customers were delighted with their rooms, and Room's importance weight for influencing overall delight with the hotel is 29 out of 100. Three more steps forward using F2 will take you to the end of the review screens.

This information can be pondered to choose possible areas for quality improvements, as described earlier. Hard copies of the review screens can be obtained by pressing F7 and following the menu directions.

THE INPUT SCREENS

Pressing F5 switches us back and forth between the review screens we just examined and the part of the program that uses additional, managerially supplied inputs to compute the projected financial impact of suggested programs. F5 can be used at any time to go back to the review screens should questions about them arise. Use F5 now to go back and forth between review and input screens. End up back at the first input screen.

Customer and Market Information

The first input screen asks for information about the size and profitability of the market, which can be used to transform market share changes into revenue figures. In particular, the total market in which Cumberland Hotels competes has annual sales of 35 million stays per year (enter 35000 for number of transactions), each an average of 1.2 days in length, and is growing at 1.5 percent per year (enter 1.5 for market growth). The average person in this market stays in a hotel an average of 2.4 times per year (enter 2.4 for number of transactions), and Cumberland's contribution (revenue minus variable costs) is about $30 per visit (enter .030). Notice that the number of transactions and the profit per transaction are expressed in thousands of units. The input cells should appear as:

35000
1.50%
2.4
0.03

The model is now ready to begin calculating the profit impact on this hotel chain of various quality-improvement programs.

Attribute Under Consideration

Advance (F2) to the next screen to choose a target area for quality improvement. We will concentrate on solving some of the hotel chain's problems before we consider the benefits of delighting more of our customers. The units that are most influential on overall satisfaction, according to the review screens, are Bath, Staff, and Room, in that order. However, fully 17 percent of customers are dissatisfied with their rooms, while far fewer are dissatisfied with the other units. Let's see what we can do about eliminating dissatisfaction with rooms.

Of the Room dimensions, two stand out as simultaneously having the most problems (highest percent dissatisfied) and being the most influential on overall satisfaction with the room (highest importance weights). They are Dimen 2 (the bed) and Dimen 3 (the TV). Both have 9 percent dissatisfied customers and have importance scores of 38. Consider the TV problem. Research has found that most of the complaints with TVs are associated with the older, worn-out models. Cumberland Hotels currently has about 4,800 rooms, each with a TV, and they are replacing them on a five-year cycle—i.e., 20 percent or about 960 TVs are being replaced each year at a net cost of about $250 per set. That is an annual expenditure of $240,000 on TV sets. If the hotel were to switch to a three-year replacement cycle, it would incur an annual cost of .33 × 4800 × $250, or $400,000 per year. Is this program a reasonable expenditure of funds?

The input screen asks us to identify the service attribute that the program is likely to affect. Press F4 to enter data on this screen. Notice that there are two guides telling you what to choose. The highlighted box in the table indicates what value the program is waiting to receive, while the prompt at the top of the screen also describes the required input. Our TV replacement program will affect a specific unit, so enter the number 1 in the first space. The program will automatically go to the next input. The Room is Unit 1, so enter the number 1. The next prompt asks whether the program's effect is at the overall level of Room, perhaps simultaneously affecting several dimensions, or whether it is more targeted at one dimension. Enter a 2, since our program is focused on satisfaction with television sets only. Enter a 3 to specify the third Room dimension. The input cells should appear:

Expected Percentages of Dissatisfied Customers

Press F2 to move to the next input screen, which asks us to predict the likely effectiveness of four different levels of effort focused on the problem at hand. Two of them are extreme values, and two are intermediate values with which management should feel reasonably comfortable—for example, the current level of spending, and that of the plan under consideration. Data for this input can come from several possible sources. In the case of the TV replacement cycle, the hotel could have done a test in a limited number of hotels to see how the increased replacement schedule decreased the number of complaints. Empirical evidence of this sort is best, but not always available. Alternatively, management may have a good feel for the likely effect of the change, based on experience and an analysis of the source of the problems. (If all complaints were generated by four- and five-year-old sets, then our program could conceivably eliminate them all.) This use of managerial estimates has a proven history in management science modeling of providing valid forecasts.[1] The term for this use of managerial judgments in forecasting is "decision calculus." Note that even if empirical data were available, some managerial judgment must still be applied to account for possible competitive initiatives and other changes that may make empirical data collected in the past invalid for the future. For example, if it is known that a major competitor is about to replace all its televisions with new big-screen sets and customers' expectations about what hotels should offer are raised, then our program to replace our standard sets more often is likely to find customers still rating our TVs "worse than expected."

Press F4 to begin entering data. First, a dialog box will ask whether our program is intended to convert "unhappy" (i.e., dissatisfied) customers to satisfied ones, or to delight customers currently only satisfied. (The new term "Unhappy" was used to give the choice menu a letter different from "Delighted.") Choose "Unhappy."

The two extreme levels of spending are fixed at zero and infinity. We will supply, in turn, the two values 240 and 400, for the two levels of spending, $240,000 and $400,000, as discussed above. The cursor now moves to the effectiveness column. These are the values that must be determined based on experimental data and/or judgment. Let's assume the following values, which should be entered as numbers between 0 and 100:

■   At $0 assume 70 percent of customers would complain (i.e., enter 70)

■   At $240,000 we are experiencing 8.7 percent dissatisfaction (enter 8.7)

■   At $400,000 we expect to have only 2 percent dissatisfaction (enter 2)

■   At an infinite spending level we feel that there would be no complaints (enter 0).

The input screen should read:

$0	70.00%
240	8.70%
400	2.00%
$Infinity	0.00%

**FIGURE A1-2   THE PREDICTED RESPONSE CURVE FOR CUMBERLAND HOTELS'
TV REPLACEMENT PROGRAM**

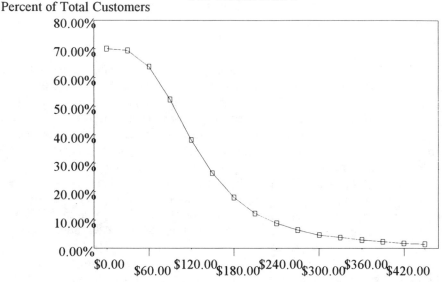

Change in Dissatisfied Customers
For Room Dim 3

Percent of Total Customers

Service Improvement Effort ($000)

If you have made a mistake in entering a number, the program will give you the option of reentering the correct values.

ROQ now fits a continuous curve (described in more detail later) to these four values, which can be used to interpolate the likely percent of dissatisfied at values other than those specified. Respond to the prompts at the top of the screen. Tell ROQ to graph the curve from a minimum value of $0 (type 0) to a maximum of $450,000 (type 450). The resulting graph, shown in Figure A1-2, shows a curve implied by the four data points we provided. These curves are not always meaningful if different programs are purchased in discrete jumps; but when spending can be increased in small increments with continuously improving results, then the intermediate values can be very useful in estimating the likely effectiveness of other levels of spending. In the case of our TV replacement program, we have a virtually continuous range of options depending upon how frequently we wish to replace television sets—from never to daily.

Touch any key. The screen reverts back to the table of values of spending and percent dissatisfied from which the graph was created. You are asked if you wish to save the graph as a *.PIC file. Try it, by typing 1 ("Yes") and typing the name of the file, TV1. The program stores it in the current directory with the name TV1.PIC. You will later be able to print the graph from DOS by typing

PICPRT TV1.PIC

The file can also be imported by most modern word processing and graphics programs. The table that now appears on the screen also shows the average retention rate for the hotel's customer base, given various levels of percent dissatisfied. This column is based upon the chain of statistical relationships estimated earlier linking percent dissatisfied with TVs to the percent dissatisfied with the room to the percent dissatisfied overall to the average retention rate. ROQ will use these values and interpolated intermediate values to estimate the change in switching patterns and market share that will result from different levels of spending on the quality program.

Profit Impact of Service Quality-Improvement Effort

Press F2 to move to the next screen. This input screen asks for some cost and savings data associated with the program so that NPVs of various levels of spending can be calculated. Press F4 to begin entering data.

The first input asks about cost savings. Many quality programs not only generate higher loyalty among customers but can also save costs by reducing waste and rework and by introducing efficiencies, as described in Chapters 6 and 7. ROQ assumes that all cost savings cash flows are realized at the ends of periods for the purposes of net present value calculations. The first inputs ask about the possible cost savings, their magnitude, and the periods in which these savings will be realized. As it happens, our TV replacement program offers no such efficiencies, so enter a zero (0) in the first box. ROQ will realize that the next two questions about the duration of cost savings are meaningless and proceed to the fourth box, which asks about the timing of expenditures.

Expenditures on quality can occur immediately, so the timing options allow for spending a period in advance of the realization of savings. We assume that the expenditures on TVs will start immediately at the beginning of the first period, so enter 0 as the First Period for Expenditures. The model allows for up to a 10-period horizon to allow effects to fully develop. Even though we may expect this program to become obsolete well before 10 years is up, let's use the full capacity of the ROQ model. Enter 10 for the Last Period for Expenses.

Our program intends to replace TVs every three years, so perhaps three years is an appropriate horizon for computing the NPV. One could also argue for longer or shorter periods, depending upon the firm's investment criteria. Enter 3 for the number of periods. Finally, assume the firm currently uses a 15 percent hurdle rate (enter 15) for evaluating investments. The cells of the input screen should appear as:

0
1
10
0
10
3
15.00%

The program is now ready to begin computing the net present value (over a three-year horizon at 15 percent) of various levels of spending on TV replacement, based upon the relationships in the customer and market switching surveys and on the data supplied by us in the input screens.

ROQ automatically moves you to a screen summarizing all relevant facts about our current investigation, and asks if all the data are correct. If you indicate that something is not correct, you are put back in search mode. That means you can use the F3 key to go back to the input screen you wish to change, and then return (via the F2 key) to this summary screen. Pressing F4 will allow you to resume computing NPV values.

If all is now correct, ROQ asks you for the minimum and maximum levels of spending per period over which you would like it to compute the NPV. Enter $0 (i.e., 0) and $600,000 (type 600) to span the set of alternatives we have discussed. The program will divide the range into 15 equal increments and compute the NPV at each break point. As soon as it is finished, a graph of the results appears on the screen, as in Figure A1-3. Note the units on the vertical axis. ROQ always presents NPV in thousands ($000), but because of the magnitude of this problem, the units are expressed in thousands yet again. Therefore, the vertical axis should be read as millions of dollars.

The vertical axis on this graph requires explanation, as it carries more information than just the net effect of the quality spending program. Basically, one uses the curve to compare the profitability of different programs by looking at the difference between their values on the vertical axis, not by reading their absolute levels.

The NPV calculation reflects deviations from the current level of sales per period. It therefore reflects the net effect of several forces in the market beyond just the level of spending. The market may be growing (in this case at an annual rate of 1.5 percent), and market share may be changing due to many factors, including our attractiveness to new customers, the retention rate of our competitors, and our own retention rate. Only the last of these is related to the level of spending in the model,

**FIGURE A1-3   THE NPV FOR VARIOUS LEVELS OF SPENDING
ON TV REPLACEMENT**

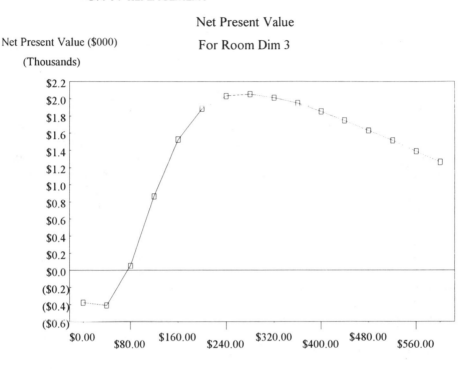

Net Present Value

Net Present Value ($000)   For Room Dim 3

(Thousands)

Service Improvement Effort ($000)

and it may not have a strong enough effect to reverse a decline or rise in share if other factors are strong enough. If a market is growing fast enough, it is possible for sales to go up, even if we lose share. If our competitors have better retention and attractiveness than we do, it is possible that we will lose share even if we start to spend heavily on quality—although we would presumably lose share more slowly than if we neglected quality.

The point is that retention is only one of many forces causing our sales to drift upward or downward over time. The NPV calculation computes the net value of all of those factors. To compare programs, we must look at the difference that our programs make to that drift, rather than whether the actual drift is up or down. Therefore, the proper way to interpret Figure A1-3 is by finding its height at the current level of spending and noting the vertical difference in dollars between the curve at that point and at other points of interest.

For example, in Figure A1-3 at the $0 spending level on TV replacement, the graph shows a net loss of $381,000 from current sales levels. The graph does not say that Cumberland Hotels would lose money if no money were spent on TVs. It says that the net present value of revenues over the next three years would be $381,000 lower than a baseline value—namely, what they would be if sales remained at their current level. This value is the net effect of several forces. We will see later that market share is predicted to drop if spending programs for replacing TVs are dropped altogether. According to the graph, this drop will be significant enough to offset two positive forces: the market growth rate over the next three years and the cost savings from not replacing TV sets.

We can compare this value to that at the current level of spending, $240,000. According to the graph, the discounted net effect over three years of the current program will be $2,021,000. Again, this is the net effect of the current share drift, market growth, and the continued level of spending. It is $2,261,000 higher than the NPV of the $0 (no spending) plan. In fact, it appears to be nearly optimal. The maximum return appears to be at a level of spending near $280,000, just above our current level of spending. Beyond $280,000, the graph indicates that we are beyond the point of diminishing returns, and the NPV of more expensive programs starts to drop.

The proposed level of spending, $400,000, does not appear especially attractive. The curve indicates that the NPV would be $176,000 lower than the predicted value of the current program. The additional sales generated through higher retention would evidently not be enough to offset the increased expenditures on televisions. It appears that Cumberland Hotels must find a more efficient way to improve customer satisfaction with its television sets, if that is indeed where it wishes to concentrate its efforts.

A Word of Caution

The values generated by a model are only as good as its assumptions and the data fed into it. The graph shows us the optimal value for the model—which is not necessarily the same as Cumberland Hotels' real problem. We should not, for example, recommend that they increase their replacement cycle a bit more so that it coincides precisely with the optimum value of the curve. It appears that Cumberland is pretty close to the correct value, and that, barring other reasons for doing so, the proposed more frequent schedule will not lead to a large financial payoff. But remember this warning:

Decision support systems supply input to a manager's decision.
They don't make the decision.

There may be many other factors and bits of conflicting evidence that must be weighed before arriving at the final decision.

Key variables should be submitted to sensitivity analysis—i.e., tested by trying other values to see how sensitive the model is to variation in these variables. If its results are found to vary wildly when some of its inputs are changed slightly, then it becomes incumbent upon management to spend money on research to narrow the estimates of those sensitive variables. We will suggest some sensitivity analysis for this problem in the exercises at the end of the appendix.

Market Share Impact of Quality-Improvement Effort

If the company is currently more interested in building up its share than focusing on profits, ROQ can also forecast the firm's market share over time due to various levels of the TV replacement program. Press F2 again to take you to the last input screen. Press F4 to enter several levels of spending. The first input is the number of different specific spending levels we would like to examine. The model can accommodate up to four, so let's use the full capacity of the model. Enter 4. Four boxes will appear, and you will be asked to enter the four levels of spending you wish to explore.

Let's examine four values that span the programs under discussion: $0 (enter 0), $240,000 (enter 240), $400,000 (enter 400), and $600,000 (enter 600). The cells of the input screen appear as:

4

0
240
400
600

The resulting graph is in Figure A1-4. Note the different trajectories that market share is projected to take over time, given the different effects on retention rate that the corresponding levels of dissatisfaction are expected to generate, all starting from the hotel chain's current share of 2.52 percent. At the current level of spending ($240,000), market share is expected to rise to a new level of 2.57 percent. At the level of spending near that of the proposed new program, $400,000, the share is predicted to be somewhat higher, but evidently not enough to offset the additional annual spending. It is evident that $400,000 is past the point of diminishing returns in that the curve at $600,000 is almost identical with it. The lowest trajectory is that at $0 spending. The sharp drop explains the loss of revenue even with a savings of $240,000 per year. Touch any key to see the table of values on which the market share graph is based.

All of the input screens used to analyze the replacement program, including the tables of figures used to produce the graphs, can be printed by pressing F7 and choosing the Input Screens option.

**FIGURE A1-4   PROJECTED MARKET SHARE TRAJECTORIES FOR FOUR TV
REPLACEMENT SCHEDULES**

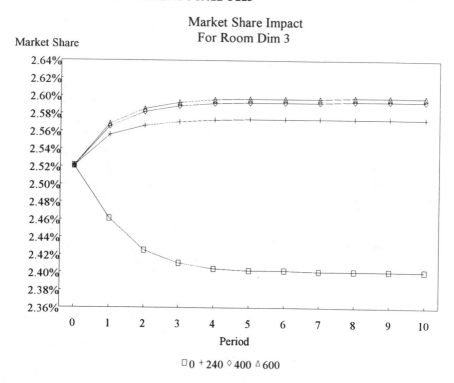

Market Share Impact
For Room Dim 3

Performing Sensitivity Analysis

Sensitivity analysis is easy to do with the ROQ. All of the inputs can be easily
altered by going back to the appropriate screen, reentering the values including the
new value for the variable being tested, and then returning to the screen from which
the output variables of interest (either NPV or market share) are computed. At that
point, pressing F4 will begin the new calculations. For example, NPV calculations
are very sensitive to both the horizon over which they are computed and the interest
rate. Go back to the screen titled "PROFIT IMPACT OF SERVICE QUALITY
IMPROVEMENT EFFORT" and change the number of periods to one year. Then
go forward to the NPV screen, press F4, and recompute the NPV from $0 to
$600,000. Stay at that screen and narrow down the location of the optimal spending
level by having ROQ graph the curve over a smaller interval around the high point
on the curve. Try other input values as well. What would the implication be if a
spending level of $0 produced only 50 percent dissatisfied?

USING DISCRETE PROGRAMS

Press F5 to go back to the review screens and find the data on delighted customers for Restaurants. The most important dimension, Dimen 3, is the variety offered to restaurant diners. Of the customers responding, 29 percent are delighted with the variety being offered and, according to the screen describing dissatisfaction, 12 percent are dissatisfied. That leaves 59 percent of the customers who are satisfied and could be delighted by further additions to the menu. Again, we will assume that the kinds of service enhancements needed to delight customers—unusual, unexpected ethnic offerings, perhaps—will not necessarily satisfy those customers who find the current service inadequate to even meet their basic expectations. Satisfying the disgruntled customers may require a wider array of more standard items rather than fancy, unusual dishes.

Expanding the menu to contain a set of special dishes thought to offer special appeal to the hotel's target market will involve a capital outlay of $50,000 per hotel, mostly for improved cold storage to keep unusual, infrequently asked for ingredients. It is anticipated that this new capacity to increase variety will increase the percent of delighted customers to about 45 percent. (Again, this figure could be based on test marketing the new menu or on managerial judgment.) Assuming a constant 24 hotels in the chain during the next few years amounts to an initial outlay of $1,200,000. The new system will also reduce the cost of other foods lost to spoilage, outweighing the cost of running the equipment, for an estimated net savings of $86,400 per year throughout the hotel chain. Assume a five-year horizon and the chain's standard 15 percent discount rate. Will this investment be worth the money?

This example is very different from the first one we considered. Here we are not seeking an optimal spending level along a continuum, but rather are comparing two very different discrete alternatives. To determine their relative value, we will use the ROQ model twice—once to compute a baseline NPV for the current program, which involves no unusual spending or savings, and then again to compute the marginal change in NPV for the proposed new program, which contains a one-time capital outlay and periodic savings.

Go back to the ATTRIBUTE UNDER CONSIDERATION screen, press F4 to enter data, and select Unit 6 (Restaurant) and Dimension 3. (The numbers entered should be 1, 6, 2, 3, respectively.)

Computing a Baseline NPV for the Current Program

On the next screen, we must establish in the ROQ model a linkage between our baseline program and the current retention rate for purposes of computing NPV. It is unnecessary to sort through accounting data to establish the actual current spending on restaurant variety, because we will only be looking at marginal change—i.e., what difference the additional spending will make to contribution. So we can set the current baseline value at any number and consider the new program to simply be a $1,200,000 increase above it. For simplicity, let's choose $0 as the baseline spending value. Press F4 and choose "Delighted" as the task option. The input screen requires four levels of spending, although we are really interested only in what the model computes for an NPV at $0. So enter any two intermediate spending values you wish, say 1 and 2. For percents of delighted at the various levels, we must be certain that the current level of delighted corresponds to the baseline spending level, so enter 29.4 opposite $0. Again, we can enter any arbitrary values you wish for the remaining three entries. However, ROQ requires that additional spending have a positive influence on percent delighted, so the numbers must increase. For example, enter 40, 50, and 60. The recommended cells of the data entry screen are:

$0	29.4
1	40
2	50
$Infinity	60

The resulting graph, over any range you select, say 0 through 60, is as arbitrary as the numbers entered, since it does not correspond to levels of actual programs.

At the PROFIT IMPACT screen, we must describe the current program. Cost savings are $0, and the range of periods is from period 0 to period 10, because we wish to see the consequences of using the current program into the future. Choose five periods and an interest rate of 15 percent. The data entry cells should be:

0
1
10
0
10
5
15

Again, we are only interested in the NPV of this option when spending is $0, so set the lower range of the graph at 0, and choose any arbitrary upper limit, say 15. The program computes the NPV of the baseline program as $5,393,000. This is the value to which we will compare the proposed new program. Don't bother to examine the graph or the table of values beyond the $LEVEL = $0.00. Those values are meaningless for our problem.

Determining the Value of the New Program

Return to the screen in which we entered the function describing customer response to the program, press F4, and choose "Delighted." We must now link the proposed level of spending to its anticipated effect on customer retention. Enter our proposed capital outlay, 1200, as the first intermediate value. We are not interested in a second program, so enter any arbitrary higher value, say 3000, as the second intermediate spending level. For the levels of satisfied customers, the only value that we definitely must specify is the anticipated 45 percent at 1200. However, we will be able to make an interesting observation later if we also associate $0 spending on this screen with the baseline level of customer delight. So, enter 29.4, 45, and any two other increasing values, say 60 and 70, in the percent DEL column, as shown:

$0	29.4
1200	45
3000	60
$Infinity	70

Again, the graph is not meaningful, so choose any limits, say 0 and 3000, to move on to the next screen.

At the PROFIT IMPACT screen, we must describe the proposed program. Expected annual cost savings are to be $86,400 (enter 86.4), and will extend over the life of the program from periods 1 through 10. (Since we are planning only a five-year NPV, one could actually enter any number from 5 through 10 as the upper limit.) The storage units will be paid for up-front, so the first and last periods for this expense are period 0. (We're more interested here in cash flows than in accounting profits using depreciation schedules, etc. We assume the entire expense takes place immediately.) The number of periods is 5 and the interest rate is 15 percent. The input cells should be:

86.4
1
10
0
0
5
15

Because we are interested only in the NPV of this single program, the range of calculations is arbitrary, as long as $1,200,000 is one of the breakpoints of the calculation. So choose 1200 as the lower limit, and any higher value as the upper limit. ROQ computes an NPV of $4,493,000 for this program, $900,000 lower than the NPV of the baseline program. This strongly indicates that the program is not worth the investment. The difference in retention rates between satisfied and delighted customers is simply not great enough, and this attribute is not influential enough to make a sufficiently large change to retention rates. Again, it should be noted that this comparison is subject to the many assumptions in the model, and should be used with other managerial considerations in making this capital decision.

How ROQ Treats One-Time Investments

A comment about the model's incorporation of one-time investments is in order. For example, in this case we have told the model that we will be spending enough to increase the proportion delighted to 45 percent in the initial period, and that we will not be spending any additional money in subsequent periods. However, ROQ will assume that the positive impact on delight and retention of this investment will continue throughout the analysis period. It uses the 45 percent figure for the rest of the horizon, and does not reduce the percent delighted back to the figure associated with $0 spending in the subsequent periods. The reduction in spending affects only the cash flows used in computing NPV.

Finally, use F2 to go to the MARKET SHARE IMPACT screen. We can now compare the market share impacts of the two programs, because when we entered the most recent response function, we indicated that $0 spending is associated with the current percent delighted of 29.4 percent, and that the new program associated with the $1,200,000 outlay produces 45 percent delighted. The market share calculation is based only on the implied retention rates associated with different spending programs, not on the timing or nature of their cash flows, so we can compare the market share effects over time of these two very different spending programs. Choose two spending options, 0 and 1750. The resulting graph shows the market share trajectories for the two programs both rising from 2.52 percent to 2.57 percent. Notice that they are identical within round-off error. The higher contribution estimated for the proposed new system is due almost entirely to cost savings rather than higher repeat business.

In fact, delighting customers with restaurant variety has very little impact on retention rates. (Go back to the response function review screen to see how nearly constant retention levels are for the different percentages of delighted customers, rising from 65 percent to only 65.01 percent.) There are two reasons for this weak effect. First, delighted customers do not have a much greater return probability than satisfied ones, and second, the ripple effect of delight with the variety dimension on up to overall delight with the hotel is ultimately rather small.

These two examples have been intended to introduce you to the ROQ model and its use as a managerial tool, and to begin to stimulate thinking about how to measure the payoff from quality programs. Proficiency with the model will only come with further use. Explore the model to test its sensitivity to various assumptions. The cases later in the book will give you further experience with using the model as a decision tool.

TECHNICAL DETAILS OF THE ROQ MODEL

Remember the warning about decision models posted above. Decision support systems are tools that provide useful input to a manager's decision, but their findings must be considered along with many other objective and subjective factors. They seldom provide definitive answers by themselves. This is because the outputs of any model are dependent upon the model's assumptions, and all decision models are based on a simplified view of reality. Therefore, before using any decision model, it is important to have at least an intuitive understanding of how the model works. This understanding should give the user more confidence in using the model's forecasts, because he knows when they are likely to be accurate and when they are unlikely to reflect some key aspects of reality. Often, in the latter cases, the user may be able to put the appropriate "spin" on the model's output to account for the missing factors. In this section we describe several of the key assumptions underlying the ROQ model so that the user will better understand what data must be collected and how the results should be interpreted. We present them in the order in which the user encounters them.

Estimating the Importance Weights

The satisfaction importance scores are derived from regression analysis between measures of satisfaction at different levels. These measures are derived from the customer satisfaction survey by converting customer choices on the three-point scale ("worse than expected," "about as expected," and "much better than expected") into two dummy variables S and D. S is an indicator that the customer is at least satisfied:

If a customer rates an item as worse than expected, then $S = 0$.

If the item is about as expected or much better than expected, $S = 1$.

D is an indicator of delight.

If a customer rates an item as about as expected, then $D = 0$.

If the item is rated much better than expected, $D = 1$.

The overall unit importance weights are obtained by using regression methods to estimate the following relationship between the dummy variable for overall satisfaction and the satisfaction dummy variables for the units:

$$(A1.1) \qquad S_{\text{Overall}} = a_0 + a_1 S_{\text{Room}} + a_2 S_{\text{Bath}} + a_3 S_{\text{Staff}}$$
$$+ a_4 S_{\text{Service}} + a_5 S_{\text{Grounds}} + a_6 S_{\text{Restaurant}}$$

The coefficients of this equation are used as importance weights, provided they are positive. Negative coefficients are presented as zero, since we are interested only in those variables that can have a positive influence on overall satisfaction.

Customer satisfaction surveys are often plagued by the statistical problem of multicollinearity—i.e., customers tend to rate several of the units or their dimensions similarly, almost always high or low together. There are several possible reasons for this. First of all, different people have different levels of acceptance: Some may be very critical or very enthusiastic about everything, and tend to use only one end of the scale. This behavior is much less common with the three-point comparison-with-expectations scale than with the traditional five-point poor-to-excellent rating. A second cause of multicollinearity is due to "halo effects," in which delight or disappointment with some central aspect of the service tends to color customer attitudes about other aspects. So, for example, if there is a problem with the bath, the customer may be very critical of the room as well.

Both of these reasons assume that multicollinearity is a spurious phenomenon, but there is always the possibility that two variables really are correlated. This is not as likely if the units are truly independent processes of the service. However, if—perhaps due to administrative structure—two units generally perform well or poorly together, then it will be much harder to separate the effects of each unit on overall satisfaction. In that case, the data should perhaps be grouped together and the two units treated as a single unit.

Multicollinearity poses problems for the estimation of importance weights, because when scores on several units vary together, it is impossible to know which ones are relatively more important in driving the overall score. As a result, the estimates of the coefficients of the units' scores in Equation (A1.1) are subject to a great deal of uncertainty. To minimize the effect of multicollinearity and to obtain the most stable estimates possible from the information supplied, the ROQ model estimates Equation (A1.1) using a technique called the equity estimator[2] instead of ordinary least-squares methods.

The Chain of Effects from Dimension to Retention

The coefficients in Equation (A1.1) and those relating units with their dimensions are linked together to predict the effect on retention that is likely to occur given a change in satisfaction levels at the unit or dimension level. For example, from Equation (A1.1) we have the relationship

$$(A1.2) \qquad S_{\text{Overall}} = c_{\text{Room}} + a_1 S_{\text{Room}}$$

where c_{Room} is a constant. In particular, in a regression relationship it must be true that

$$(A1.3) \qquad \overline{S}_{\text{Overall}} = c_{\text{Room}} + a_1 \overline{S}_{\text{Room}}$$

where $\overline{S}_{Overall}$ and $\overline{S}_{Room}$ are the mean values of $S_{Overall}$ and S_{Room}, respectively. From relationship (A1.3) it is easy to solve for c_{Room}.

Because these are both 0/1 variables, their means equal, respectively, the percent of customers satisfied with the hotel overall and the percent satisfied with the room. That is, Equation (A1.2) is not just a relationship between individual satisfaction ratings. It relates percentages of customers satisfied with different levels. ROQ uses this relationship to predict changes in the percent of customers satisfied with the hotel overall from changes in the percent of customers satisfied with the room.

The ROQ model derives similar relationships at each dimension level, such as

(A1.4) $$S_{Room} = c_{TV} + a_{13} S_{TV} \text{ or}$$

(A1.5) $$D_{Room} = d_{TV} + b_{13} D_{TV},$$

the corresponding equation for delighted customers. By combining Equations (A1.2) and (A1.4), we have

(A1.6) $$\begin{aligned} S_{Overall} &= c_{Room} + a_1 (c_{TV} + a_{13} S_{TV}) \\ &= c'_{Room} + a' S_{TV} \end{aligned}$$

a direct relationship between the percent of customers satisfied with the TVs in their rooms and their overall satisfaction with the hotel.

To complete the linkage to retention, ROQ determines the average retention rates for each group—those dissatisfied, those satisfied, and those delighted with the overall service. If we call them R_1, R_2, and R_3, respectively, then the mean retention rate among the hotel chain's customers is

(A1.7) $$\overline{R} = (1 - \overline{S})R_1 + \overline{S}R_2 + \overline{D}(R_3 - R_2)$$

where $\overline{S}$ is the mean value of S, or the percent of customers who are at least satisfied overall, and $\overline{D}$ is the mean value of D, or the percent of customers who are delighted with the hotel overall.

This completes the chain of equations. For example, Equation (A1.6) shows how a change in percent satisfied with a dimension affects the percent satisfied overall, and (A1.7) shows how these changes in the percent satisfied overall are translated into new average retention figures.

The Response Function

The curve fit to the data on program effectiveness is often called the ADBUDG curve, because it was used in one of the original decision calculus models dealing with advertising spending,[3] and has been used in many other decision calculus applications since then. The curve for delight has an equation of the form

(A1.8) $$Y = Y_0 + (Y_1 - Y_0)\frac{X^\alpha}{\gamma + X^\alpha}$$

where Y_0 is the percent delighted at $X = 0$ spending, Y_1 is the upper limit of delight at infinity, and α and γ are positive constants that control the shape of the curve to make it pass through all three finite points and properly approach the upper limit at infinity. The shape of the curve is flexible enough that, by properly choosing the four parameters, the curve can be made to pass through any three sequentially rising values. The formula for falling percentages of satisfaction with increasing spending is a variant of (A1.8), which can also be fit to any three falling points and a somewhat lower limiting value:

$$Y = \left(Y_0 - Y_1\right)\left(1 - \frac{X^\alpha}{\gamma + X^\alpha}\right) + Y_1$$

where Y_0 is the maximum percent dissatisfied when spending $X = 0$, Y_1 is the lower limit of dissatisfied with infinite spending, and α and γ are positive-valued shape parameters as before.

The NPV Calculation

The net present value calculation for a particular spending pattern measures not the absolute NPV of all sales, but the NPV of the change in sales from the current level, regardless of the source of the change. As was explained above, some of the "drift" in sales may be due to better or worse retention rates by our competitors or differences in the relative attractiveness of the various firms in the market. Therefore, sales may rise or fall regardless of what we do to improve restaurant variety, for example. What is important in comparing quality programs, then, is their relative effectiveness. A program may be unable to reverse a drop in market share, but if it can slow that drop more cost-effectively than other programs, it is worth considering. In particular, the formula computes the net sales for the nth period out as sales in the nth period due to market growth and change in market share minus the current number of sales per period:

(A1.9) $\qquad \Delta S_n = MS_n(N_0 T)(1 + g)^n - MS_0(N_0 T)$

where MS_n = market share in period n

$\qquad MS_0$ = current market share

$\qquad N_0$ = current market size = number of individuals

$\qquad T$ = number of sales per individual per period

$\qquad g$ = market growth rate

The NPV of cash flows in the nth period is then computed as

$$NPV_n = (1 + r)^{-n}(p\Delta S_n - K_n + R_n)$$

where r = firm's discount rate

         p = individual contribution per purchase

         K_n = cash outlays for the quality program in period n

         R_n = savings due to quality program in period n

Computing Market Share Over Time

The focus of the ROQ model is on the improvement in sales due to improved customer retention rates. Higher retention rates cause higher sales, as defections slow market share increases, all else being equal. Therefore, the key to the ROQ calculations is the projection of market shares in future periods as a function of retention rate. The ROQ model bootstraps market share formulas forward one step at a time, from period t to period t + 1, using the information gathered beforehand and that supplied in the input screens. First some notation. Let

N_t = market size in period t

$G = (1 + g)$ = market growth factor

MS_t, MS_{t+1} = market share in periods t and t + 1

c = churn or percent of customers who leave the market each period

r_t = percent of customers retained by our firm after period t

r'_t = competitors' collective retention rate

A_{t+1} = our relative attractiveness to new customers in period t—i.e., the percent of new customers in the market who choose us.

Then the number of new customers in period t is:

(A1.10) $r_t\,MS_t\,N_t$ +   (retained customers)

              $(1 - r't - c)\,(1 - MS_t)\,N_t$ +   (switchers)

                   $A_{t+1}\,(G\,N_t - (1 - c)\,N_t)$ (new customers)

The total number of new customers in the market in period t + 1 will be $G\,N_t$, so our share will be the value in (A1.10) divided by $G\,N_t$, or:

(A1.11) $MS_{t+1} = [r_t\,MS_t + (1 - r'_t - c)\,(1 - m_t) + A_{t+1}\,(G - 1 + c)]\,/\,G$

In the model, all parameters in (A1.11) are assumed fixed for the near term except retention, r_t, which is assumed to be a function of spending on quality, as described in Equation (A1.7). While historical measures may provide some guidance about the values of these other parameters, an important role of managerial judgment is to adjust such key variables as market growth, competitive retention rates, and relative attractiveness to reflect the most likely values over the planning horizon.

CONCLUSIONS

This appendix has shown how to use the ROQ software and has explained the technical details of how the model is put together. The next chapter will further explore the case of Cumberland Hotels. This case will demonstrate how the ROQ decision support system can assist managers in making quality-improvement decisions.

NOTES

1. For example, see John D.C. Little and Leonard M. Lodish (1969), "A Medical Planning Calculus." *Operations Research* 17 (January-February), pp. 1-35; John D.C. Little (1970), "Models and Managers: The Concept of a Decision Calculus." *Management Science* 16 (April), pp. B466-B485; Robert C. Blattberg and Stephen J. Hoch (1990), "Database Models and Managerial Intuition: 50 percent Model + 50 percent Manager." *Marketing Science* 36 (August), pp. 887-899.
2. See Arvind Rangaswamy and Lakshman Krishnamurthy (1991), "Response Function Estimation Using the Equity Estimator." *Journal of Marketing Research* 28 (February), pp. 72-83; and Lakshman Krishnamurthy and Arvind Rangaswamy (1987), "The Equity Estimator for Marketing Research." *Marketing Science* 6 (Fall), pp. 336-357.
3. John D. C. Little, op. cit.

CASE: CUMBERLAND HOTELS

BACKGROUND

It was June of 1987, and John McCormick, CEO of the Cumberland Hotels chain, was reviewing the results of the company's guest satisfaction surveys. John was hoping that the research would point to areas in which the company could build customer loyalty through improved service.

John had been CEO for just under six months. Prior to joining Cumberland Hotels, he had been the chief operating officer (COO) for Deluxe Suites, a national hotel chain. Cumberland had actively recruited John to take over for the retiring founder and CEO, Cook Welch. Welch was in his seventies, and felt that the company needed someone who could devote more time to the business. He would stay on as chairman but would no longer manage the activities of the company.

Welch personally selected John to take over the management of Cumberland Hotels because he had built a reputation as an innovative turnaround specialist at Deluxe. While Cumberland was doing reasonably well, Welch believed that the company had grown complacent, resulting in little change in market share over the previous three years.

CUMBERLAND HOTELS

Cumberland Hotels began in 1959, when Welch built his first hotel outside of Nashville, Tennessee. He slowly began building additional properties, ultimately creating a hotel chain consisting of 24 properties located in five southeastern states.

Cumberland Hotels focused on the business traveler. The strategy was to provide guests with comfortably furnished, moderately priced rooms. Further, consistent quality and appearance was demanded of all the properties.

Welch realized that the success of his business depended on the quality of each hotel's general manager. Therefore, he provided Cumberland managers with extensive training before they were allowed to manage a property. Further, he required that all managers meet annually for a week of additional training. These annual meetings were also structured so that managers could talk with one another about successful and unsuccessful programs.

Welch was a very hands-on manager. He made it a point to stay at each of his properties at least once every six months. Because these visits were always unannounced, he believed that general managers would make a special effort to stay on top of their operations.

During the mid-1980s, Welch's health began to make the frequent travel more difficult. As a result, he believed that a new CEO was necessary if the company was to remain competitive. The company had always been a family-run business; however, Welch believed the company had grown complacent. Therefore, he decided to select an outsider as CEO, surmising that Cumberland Hotels would have the advantage of new ideas while he would be able to maintain oversight regarding the transition.

Although his son, Anthony, was president of Cumberland Hotels, Welch believed that he was not yet ready to be CEO. Anthony was extremely unhappy with this decision and threatened to resign. Welch was ultimately able to change his mind by explaining why an outsider was needed and by making it clear that his wealth was tied to the long-term viability of the company. (The Welch family was the sole owner of the Cumberland Hotel chain.)

Welch then began a search for his replacement as CEO. This search led him to John McCormick, COO of Deluxe Suites. At the time, John had been COO at Deluxe for about four years. During that time, Deluxe had gone from a company that was consistently losing market share and money to a growing and profitable enterprise. John was widely credited for this turnaround.

John's strategy at Deluxe was to simultaneously cut costs and improve quality at the 136 Deluxe Suites located throughout the United States. He initiated a quality-improvement program that focused on meeting the expectations of customers. To gain insight into customers' needs, he conducted regular research on guests. Further, to make sure that the hotel managers were serious about customer satisfaction, part of their compensation was based on how guests rated their experience at their hotel.

Welch believed that John's approach would work well at Cumberland Hotels. John, however, did not immediately accept Welch's offer. He was already doing well at a job he loved. Further, he realized that Cumberland Hotels was a family business and that he would be grooming Anthony Welch to ultimately take over as CEO. Cook Welch, however, was certain that John was the man for the job and asked that he take more time to consider the offer.

John then began research into Cumberland Hotels. Deluxe and Cumberland were not in the same market, and therefore did not compete for the same customers. Hotels in Cumberland's market accounted for approximately 42 million nights per year. The average stay per guest is 1.2 nights, and the average customer stays in a hotel 2.4 times per year. The market was growing at about 1.5 percent per year; however, Cumberland's share of this market was a relatively small 2.52 percent.

Cumberland Hotels also averaged only 200 rooms per hotel, versus an average of 600 rooms for a Deluxe Suites hotel. Occupancy averaged just over 60 percent per night. Because room rates were moderately priced, the average room contributed approximately $25 in profit per night to the hotel.

John's review of Cumberland Hotels led him to believe that there was both substantial opportunity and risk to Welch's offer. On one hand, he had been offered a stake in the company should certain profit objectives be reached, but unlike Deluxe Suites, he would be dealing with a lower-margin property and a customer base that may not have the same expectations of service. After considerable deliberation, John decided to accept the offer if two conditions were met. First, that he be given adequate time to allow his programs to work, and second, that he be allowed to manage the business as he saw fit (with the exception of standard boardroom oversight). Welch agreed.

THE TRANSITION

Immediately after taking over as CEO, John went through an abbreviated version of the training program required of all general managers in order to gain insight into the company's standard procedures. He then visited each Cumberland Hotel and met with the general manager. While at the hotels, he made a point of speaking informally with guests to uncover their opinions and suggestions.

After visiting each of the properties, John determined that the hotels were doing a reasonably good job of serving their client base. His talks with guests led him to believe that they considered Cumberland Hotels to be good, not great hotels. They chose to stay at Cumberland because they believed that the trade-off of amenities for price was reasonable.

Meeting with Cumberland's customers convinced John of the need to conduct regular market research. He believed that Cumberland's customers were extremely value-conscious, wanting comfort but not necessarily an abundance of features. As a result, he wanted to determine what factors drove customer satisfaction with the hotels so that any changes he implemented would be valued.

GUEST SATISFACTION SURVEY

Drawing on his experience at Deluxe Suites, John designed and implemented a Guest Satisfaction Survey program. The surveys were to be administered each month and were designed to measure customer satisfaction with six overall processes of Cumberland Hotels' operations: (1) Room, (2) Bath, (3) Staff, (4) Services, (5) Grounds, and (6) Restaurant. Table A2-1 shows the areas measured by the survey.

The surveys were sent to a randomly selected number of guests who had recently stayed at one of the hotels. Guests were asked to answer questions regarding their most recent stay. Because ultimately John hoped to use the results of the surveys as a factor in determining management compensation, surveys were administered so that the number of respondents was the same for each hotel.

Survey Results

The results of the survey revealed that the majority of guests were satisfied with Cumberland's overall level of service. Further, the research revealed that the most important factor in determining customers' overall satisfaction was their satisfaction with the bathroom. Satisfaction with the staff and room were also found to be important. Figures A2-1 and A2-2 show the percentage of customers who were dissatisfied and delighted with Cumberland Hotels regarding various aspects of the company's service.

John's experiences at Deluxe Suites led him to believe that the initial phases of a hotel quality-improvement program should focus on eliminating problems. To him, this meant minimizing dissatisfied customers before attempting to exceed customers' expectations.

John believed that reducing dissatisfaction with the bathroom could represent an opportunity. Nine percent of the guests were dissatisfied with the bathroom overall. Of the factors determining satisfaction with the bathroom, supplies was found to be most important. Thirteen percent of customers were dissatisfied with the bathroom supplies. Currently, bathroom supplies (soap, tissue, etc.) averaged 75 cents per room per night stayed. John believed that by adding certain supplies, such as shampoo, lotion, etc., the percentage of dissatisfied customers could be reduced to 5 percent. These supplies would add 40 cents to the cost of supplies per room.

TABLE A2-1   AREAS EXAMINED BY MARKET RESEARCH

	Room	Bath	Staff	Services	Grounds	Rest
Dimension 1	Carpet	Tub	Knowledge	Wake-up Call	Landscaping	Service
Dimension 2	Bed	Cleanliness	Friendliness	Messages	Lighting	Cleanliness
Dimension 3	TV	Supplies	Professional	Check-in/out		Variety
Dimension 4	Workspace	Vanity				Quality
Dimension 5	Lighting					

**FIGURE A2-1   PERCENTAGES OF DISSATISFIED CUSTOMERS
AND IMPORTANCE TO RETENTION**

PERCENTAGES OF DISSATISFIED CUSTOMERS & IMPORTANCE TO RETENTION
(Management Units 1-5, Dimensions 1-10)

	Room		Bath		Staff		Services		Grounds	
	DIS	IMP	DIS	IMP	DIS	IMP	DIS	IMP	DIS	IMP
Overall	17%	31	9%	52	4%	35	9%	21	26%	0
Dimen 1	10%	0	4%	24	4%	0	5%	54	11%	50
Dimen 2	9%	38	4%	32	4%	59	12%	33	21%	0
Dimen 3	9%	38	13%	16	0%	0	4%	0		
Dimen 4	6%	23	9%	0						
Dimen 5	4%	25								

PERCENTAGES OF DISSATISFIED CUSTOMERS & IMPORTANCE TO RETENTION
(Management Units 6-9 + Product, Dimensions 1-10)

	Rest									
	DIS	IMP	DIS	IMP	DIS	IMP	DIS	IMP	DIS	IMP
Overall	24%	25								
Dimen 1	6%	18								
Dimen 2	0%	0								
Dimen 3	12%	31								
Dimen 4	18%	24								
Dimen 5										

**FIGURE A2-2   PERCENTAGES OF DELIGHTED CUSTOMERS
AND IMPORTANCE TO RETENTION**

PERCENTAGES OF DELIGHTED CUSTOMERS & IMPORTANCE TO RETENTION (Management Units 1-5, Dimensions 1-10)									
Room		Bath		Staff		Services		Grounds	
DEL	IMP	DEL	IMP	DEL	IMP	DEL	IMP	DEL	IMP
Overall 17%	29	17%	48	44%	32	22%	19	11%	0
Dimen 1 0%	0	17%	25	26%	0	47%	47	5%	82
Dimen 2 17%	37	22%	34	44%	58	41%	29	21%	0
Dimen 3 26%	38	30%	17	35%	0	13%	0		
Dimen 4 33%	22	17%	0						
Dimen 5 30%	25								

PERCENTAGES OF DELIGHTED CUSTOMERS & IMPORTANCE TO RETENTION (Management Units 6-9 + Product, Dimensions 1-10)									
Rest ·									
DEL	IMP	DEL	IMP	DEL	IMP	DEL	IMP	DEL	IMP
Overall 41%	24								
Dimen 1 12%	20								
Dimen 2 29%	0								
Dimen 3 29%	33								
Dimen 4 24%	25								
Dimen 5									

There was one piece of information on the survey that John did not expect to find. Over 23 percent of the respondents to the survey were dissatisfied with the restaurant overall. While this was not the most important factor in determining guests' overall satisfaction with the hotel, John believed that it could represent an area where improvement would be easier to achieve given the relatively large percentage of dissatisfied customers.

When the surveys were broken out for each hotel, the results showed that for one hotel, the percentage of guests who were dissatisfied with the restaurant was only 5 percent. For the remaining hotels, the percentages of guests who were dissatisfied with the restaurant were in the low to middle twenties.

Further analysis showed that the restaurant where only 5 percent of the customers were dissatisfied was not the standard Cumberland Hotel restaurant. Instead, it was a low-cost, southern cooking restaurant that catered to both hotel guests and the surrounding community. John believed that the remaining hotel restaurants could be converted to this format for a one-time cost of approximately $100,000 per restaurant.

CONCLUSION

John now faced a difficult decision. He realized that his first program must be successful if he was to get future acceptance from the hotel managers. While improving the bathroom supplies and converting the restaurants both appeared to present opportunities, he felt that cost considerations and the importance of success required that the company focus on only one program. The question then became, which of the two programs offered the best return to the company?

CASE QUESTIONS

1. Approximately how many transactions take place per year in the market in which Cumberland Hotels competes? Approximately how many customers are in this market?
2. Approximately how much does Cumberland Hotels spend for bathroom supplies each year? If the company begins the new supply program, what will be the cost for the year?

Note: The next questions require using the ROQ Software.

When calculating the net present value, assume a three-year time horizon and a discount rate of 15 percent.

3. Will the efforts to improve the bathroom supplies have a positive impact on profits? What about market share?
4. Will converting the restaurants have a positive impact on profits? What about market share?
5. Which program (if any) should Cumberland Hotels implement? Why?

ROQ ANALYSIS: CUMBERLAND HOTELS

The Cumberland Hotels case presents a good example of how a decision support system such as the ROQ software can assist managers in making resource allocation decisions regarding quality improvement. In the case of Cumberland Hotels, the CEO, John McCormick, must determine what, if any, quality-improvement steps the company can take to improve the hotel's profitability and market share. While he has the support of the company's chairman, he realizes that if he is to get continued buy-in from the individual hotel managers, he must be able to demonstrate that the program will produce tangible benefits to the company.

Research on the company's customers revealed two areas where Mr. McCormick believes an opportunity exists for significantly improving customer satisfaction (in this case, reducing the number of dissatisfied customers). The areas under consideration are bathroom supplies and the restaurant overall. McCormick believes, however, that cost considerations and the importance of the program's success dictate that only one program be selected.

CUSTOMER AND MARKET INFORMATION

Before ROQ can estimate the profit impact of a particular quality-improvement program, information regarding the firm's market and customers must be calculated and input into the model. The first thing that must be noted is that Cumberland Hotels is a transaction-oriented business. This means that the company's interactions with its customers are single events, and risk of losing a customer exists each time a customer selects a hotel. As a result, the ROQ software requires that market information be input in terms of the number of transactions. Figure A3-1 shows the customer and market information required by the model.

178

FIGURE A3-1   CUSTOMER AND MARKET INFORMATION

CUSTOMER AND MARKET INFORMATION	
Current Number of Transactions in the Market (000)	
Market Growth Rate per Period (Example: 5 for 5%)	
Average Number of Transactions per Customer (MAX=50)	
Average Dollar Profit per Transaction ($000)	

The first piece of information required by the model is the current number of transactions that take place in the market. In the Cumberland Hotels case, we are told that customers spend approximately 42 million nights each year in a hotel in this market. This, however, is not the number of transactions that take place. Instead, we need to know how many times customers selected a hotel in a year. To convert the number of nights stayed into the number of transactions each year, we must divide the number of nights stayed by the average length of a customer's stay. In this case, we are told that the length of a customer's stay is 1.2 nights on average. Therefore, the market has 42,000,000 ÷ 1.2 = 35,000,000 transactions per year.

The model also asks for the market growth rate per year and the average number of transactions per customer. Both of these figures are given in the case. Market growth is said to be 1.5 percent per year. Customers are said to stay in a hotel 2.4 times per year. (This implies that the current number of customers in the market equals 35,000,000 ÷ 2.4 = 14,583,333.)

The final piece of customer information required by the model is the average dollar profit per transaction. The case states that the average profit contribution per room each night is $25. However, the average profit figure must be *per transaction,* not per night. The easiest way to calculate this figure is to multiply the average profit per night by the average length of a customer's stay. This would give $25 × 1.2 = $30 average profit per transaction.

Figure A3-2 shows how the first input screen should look for the Cumberland Hotels case when running the ROQ software.

BATHROOM SUPPLIES

McCormick's first option was to focus on reducing dissatisfied customers regarding the supplies in the bathroom. Figure A3-3 shows the input screen in the ROQ soft-

FIGURE A3-2   CUSTOMER AND MARKET INFORMATION

CUSTOMER AND MARKET INFORMATION	
Current Number of Transactions in the Market (000)	35,000
Market Growth Rate per Period (Example: 5 for 5%)	1.5%
Average Number of Transactions per Customer (MAX=50)	2.4
Average Dollar Profit per Transaction ($000)	.03

ware that allows us to select a specific area for quality improvement. We are looking at a specific dimension within a business process (called Management Unit in the ROQ software). In this example, the management unit is the bathroom, and the specific dimension is the supplies. The case showed this to be Management Unit 2, Dimension 3. Figure A3-4 shows how the screen should look when concentrating on the bathroom supplies for the Cumberland Hotels case.

In order for the ROQ software to estimate the profit impact of improving the bathroom supplies, we must first determine how much the company currently spends each year on supplies. The case states that the company spends 75 cents per room on supplies each night stayed. Before we can calculate the total cost of supplies to the hotel, we must first know the number of rooms used by the hotel for the year.

There are two ways to calculate this figure. Because we know that guests spend 42 million nights per year in a hotel and that Cumberland Hotels has a 2.52 percent market share we can calculate the number of nights guests spent at a Cumberland Hotel. Guests spent $42,000,000 \times .0252 = 1,058,400$ nights.

An alternative method would be as follows: The case states that Cumberland Hotels has 24 properties, with an average of 200 rooms per hotel. The hotel's occupancy rate is stated to be over 60 percent. Therefore each night the hotel averages $(24 \times 200) \times .60 = 2,880$ rooms occupied. This comes out to $2,880 \times 365 = 1,051,200$ room nights per year.

The slight discrepancy in the answers for the two methods is due to rounding error. Using either number will have no material impact on the results of the ROQ analysis.

FIGURE A3-3   ATTRIBUTE UNDER CONSIDERATION

ATTRIBUTE UNDER CONSIDERATION

Firm Overall (0), Mgmt. Unit (1), or Product (2)?

Which Managerial Unit (1-9)?

Unit Overall (1) or a Specific Dimension (2)?

Which dimension (1-20)?

FIGURE A3-4   ATTRIBUTE UNDER CONSIDERATION

ATTRIBUTE UNDER CONSIDERATION

Firm Overall (0), Mgmt. Unit (1), or Product (2)?   1

Which Managerial Unit (1-9)?   2

Unit Overall (1) or a Specific Dimension (2)?   2

Which dimension (1-20)?   3

Bath

Dim 3

Because a room must be supplied after each night it is used (not each transaction) the yearly cost to the hotel of supplying the bathrooms is 1,058,400 × .75 cents = $793,800, or 1,051,200 × .75 cents = $788,400. If the alternative supply method is used, the case states that costs will increase by 40 cents per room. Therefore, the alternative costs per year will be $1,058,400 × $1.15 = $1,217,160, or $1,051,200 × $1.15 = $1,208,880.

Currently, 13 percent of the hotel's customers are dissatisfied with the bathroom supplies. McCormick believes that if the new method is used, the percentage of dissatisfied customers will drop to 5 percent.

Figure A3-5 shows the input screen in the ROQ software that allows us to estimate the percentage of dissatisfied customers given various levels of spending to improve service. In this case, however, we are only concerned with two specific spending levels, the current level and the optional level. Therefore, the percentage of dissatisfied customers that would exist if the company spent nothing or if the company spent an infinite amount of money on supplies is unimportant. The only requirement is that the percentage of dissatisfied customers at the minimum spending level be greater than 13 percent and the percentage of dissatisfied customers at the maximum spending level be less than 5 percent.

For simplicity, we will use $800,000 as the current spending level for supplies and 1,200,000 as the proposed spending level. Figure A3-6 shows an example of an acceptable ROQ input screen, where 100 percent of customers are dissatisfied at the minimum spending level and 0 percent are dissatisfied at the maximum. Figure A3-7 shows a graph of the estimated percentages of dissatisfied customers for various spending levels using the values shown in Figure A3-6.

FIGURE A3-5   EXPECTED PERCENTAGE OF DISSATISFIED CUSTOMERS

EXPECTED PERCENTAGE OF DISSATISFIED CUSTOMERS		
	$ Level	% DIS
Minimum Spending Level	$ 0	
Spending Level 1 ($000)		
Spending Level 2 ($000)		
Maximum Spending Level	$ Infinity	

13.00% are DISSATISFIED and   30.40% are DELIGHTED regarding

Bath	Dim 3

Having input the above information, all that remains is to calculate the net present value for the two supply options. Figure A3-8 shows the input screen used for calculating net present value. In this case, there is no expected cost savings from changing spending options. Expenses are expected to occur each period, so the first period for expenses is zero (0) and the last period for expenses is 10.

The case states that NPV calculations should be for three years with a discount rate of 15 percent. Figure A3-9 graphs the net present value of various spending levels from $0 to $1,500,000. This range was chosen so that the $800,000 and $1,200,000 spending levels would appear. This graph shows that the current level of spending appears optimal in terms of profit impact for the company.

Figure A3-10 graphs the market share impact of spending $800,000 and $1,200,000. The graph shows that for both spending levels, market share is expected to increase over time. The fact that market share increases but NPV declines for the $1,200,000 spending level indicates that although increasing the bathroom supplies will result in a net gain of customers, the dollars contributed by these customers will not offset the cost.

RESTAURANT

McCormick's second option was to reduce the percentage of customers who were dissatisfied with the restaurants overall. The case showed this to be Management Unit 6, Overall. Figure A3-11 shows how the ROQ input screen should look to select the restaurant overall.

FIGURE A3-6   EXPECTED PERCENTAGE OF DISSATISFIED CUSTOMERS

EXPECTED PERCENTAGE OF DISSATISFIED CUSTOMERS

	$ Level	% DIS
Minimum Spending Level	$ 0	100.00%
Spending Level 1 ($000)	800	13.00%
Spending Level 2 ($000)	1200	5.00%
Maximum Spending Level	$ Infinity	0.00%

13.00% are DISSATISFIED and   30.40% are DELIGHTED regarding

Bath	Dim 3

FIGURE A3-7   CHANGE IN DISSATISFIED CUSTOMERS FOR BATH DIM 3

Percent of Total Customers

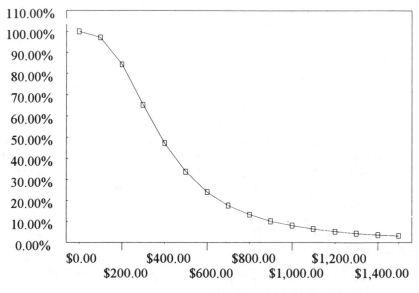

Service Improvement Effort ($000)

Before we can calculate the net present value of converting the restaurants, we must first estimate the cost of the program. McCormick estimates that it will cost approximately $100,000 per restaurant. Since there are 24 hotels, and one of the restaurants already has the desired format, the cost of the conversion is $100,000 × 23 = $2,300,000.

No information was given in the case regarding the current level of spending on the restaurants. In this situation, that information is not important. We are only concerned with determining whether converting the restaurants is a good investment decision. Therefore, we can set current spending equal to zero and set the proposed spending equal to $2,300,000.

We know that currently 23.5 percent of the hotel's customers are dissatisfied with the restaurant. With the conversion, McCormick believes that percentage will drop to 5 percent. Figure A3-12 shows an example of an acceptable ROQ input screen, where 1 percent of customers are dissatisfied at $3 million and 0 percent are dissatisfied at the maximum. The dollar amounts and their corresponding percent-

FIGURE A3-8   PROFIT IMPACT OF SERVICE QUALITY IMPROVEMENT EFFORT

PROFIT IMPACT OF SERVICE QUALITY IMPROVEMENT EFFORT	
Expected Annual Cost Savings from Effort ($000)?	0
First Period for Cost Savings (1-10)?	1
Last Period for Cost Savings (1-10)?	10
First Period for Expenses (1-10)?	0
Last Period for Expenses (0-10)?	10
Number of Periods for NPV Calculation (1-10)?	3
Interest Rate (Example: 10 for 10%)?	15.00%

FIGURE A3-9   NET PRESENT VALUE FOR BATH DIM 3

Net Present Value ($000,000)

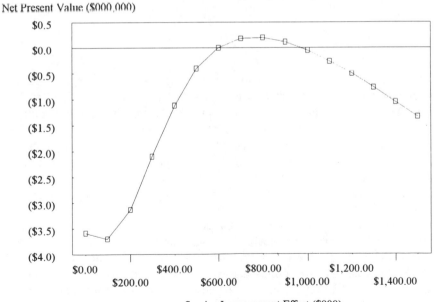

Service Improvement Effort ($000)

FIGURE A3-10   MARKET SHARE IMPACT FOR BATH DIM 3

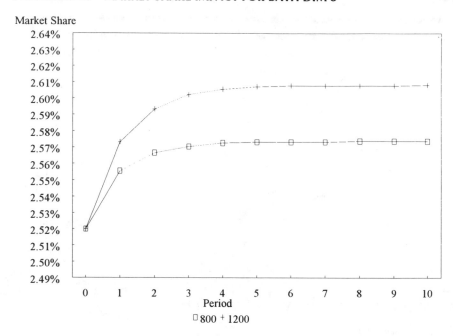

Market Share

□ 800 + 1200

ages for Spending Level 2 and maximum are unimportant; however, the percentages of dissatisfied customers must be less than the 5 percent at Spending Level 1. Figure A3-13 shows a graph of the estimated percentages of dissatisfied customers for various spending levels using the values shown in Figure A3-12.

We are now ready to compute the net present value of the current and proposed levels of spending. In this case, we are told that converting the restaurants will be a one-time cost. Figure A3-14 shows the input screen used for calculating net present value. In this case, there is no expected cost savings from changing spending options. Expenses are expected to occur only once, so the first period for expenses is zero (0) and the last period for expenses is zero (0). The case states that NPV calculations should be for three years with a discount rate of 15 percent.

Figure A3-15 graphs the net present value of various spending levels from $0 to $2,300,000. The only two points that we are interested in are the net present value at $0 and at $2,300,000. Here we find that the net present value of converting

FIGURE A3-11   ATTRIBUTE UNDER CONSIDERATION

ATTRIBUTE UNDER CONSIDERATION	
Firm Overall (0), Mgmt. Unit (1), or Product (2)?	1
Which Managerial Unit (1-9)?	6
Unit Overall (1) or a Specific Dimension (2)?	1
Which dimension (1-20)?	0

Restrnt

Overall

FIGURE A3-12   EXPECTED PERCENTAGE OF DISSATISFIED CUSTOMERS

EXPECTED PERCENTAGE OF DISSATISFIED CUSTOMERS

	$ Level	% DIS
Minimum Spending Level	$ 0	23.50%
Spending Level 1 ($000)	2300	5.00%
Spending Level 2 ($000)	3000	1.00%
Maximum Spending Level	$ Infinity	0.00%

23.50% are DISSATISFIED and   41.20% are DELIGHTED regarding

Restrnt   Overall

**FIGURE A3-13   CHANGE IN DISSATISFIED CUSTOMERS
FOR RESTAURANT OVERALL**

Percent of Total Customers

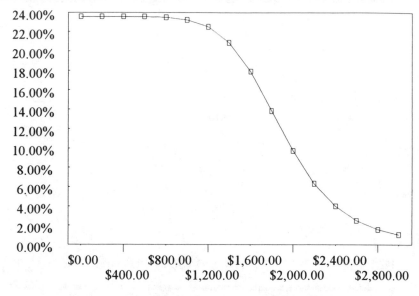

Service Improvement Effort ($000)

the restaurants is approximately $200,000 greater than keeping the restaurants' current format.

Figure A3-16 graphs the market share impact of keeping the current spending level and of spending $2,300,000 to convert the restaurants. The graph shows that for both spending levels market share is expected to increase over time; however, converting the restaurants results in a larger increase in market share than does keeping the restaurants the same.

FIGURE A3-14   PROFIT IMPACT OF SERVICE QUALITY IMPROVEMENT EFFORT

PROFIT IMPACT OF SERVICE QUALITY IMPROVEMENT EFFORT

Expected Annual Cost Savings from Effort ($000)?	0
First Period for Cost Savings (1-10)?	1
Last Period for Cost Savings (1-10)?	10
First Period for Expenses (1-10)?	0
Last Period for Expenses (0-10)?	0
Number of Periods for NPV Calculation (1-10)?	3
Interest Rate (Example: 10 for 10%)?	15.00%

FIGURE A3-15   NET PRESENT VALUE FOR RESTAURANT OVERALL

Net Present Value ($000,000)

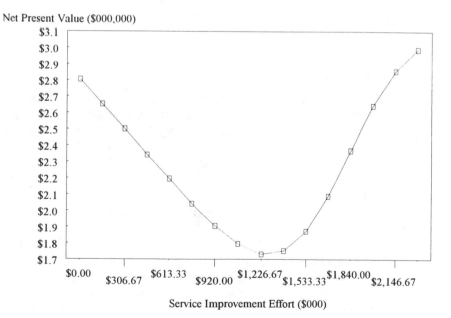

Service Improvement Effort ($000)

FIGURE A3-16   MARKET SHARE IMPACT FOR RESTAURANT OVERALL

Market Share

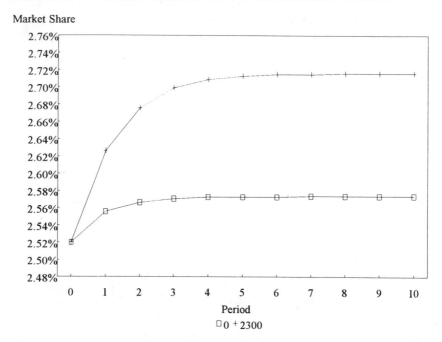

Period

□0 ⁺2300

CONCLUSION

The results of the ROQ analysis show that adding to the bathroom supplies is probably not a wise investment decision. It appears that the current level of spending is close to the optimum level. It is less clear if Cumberland Hotels should consider converting its restaurants. If one looks solely at the output from the ROQ software, converting the restaurants will generate approximately $200,000 in additional profits over three years. However, any decision support system is based on estimates and approximations. Given that there is a substantial up-front cost, and the net present values of both the current and proposed restaurants are close (relative to the cost of conversion), it can reasonably be argued that more information is needed before making a commitment. The case does state that the one restaurant that uses the proposed format caters to both hotel guests and the surrounding community. Because the ROQ model does not consider profits generated by individuals other than hotel guests, the profits from converting the restaurants could be greater than stated by the model. Without more information, however, there is no way to know if the additional patrons that will be drawn to the new restaurant will significantly affect the profitability of the investment.

CASE: ACE BANK

In January 1992, the senior officers of Ace Bank were reviewing the bank's operating results for the past year. The results were not encouraging. Although the bank had aggressively marketed its consumer deposit products to attract new customers, the bank's market share remained relatively flat at around 18 percent (actually declining slightly). Figure A4-1 shows the bank's market share for the past year and next year's projected market share. To make matters worse, the bank was experiencing high turnover among its front-line employees.

Ace Bank had actually increased its number of depositors from January 1991 to January 1992, growing by 1.76 percent for the year. However, the market had grown from approximately 196,000 depositors to around 200,000 during this same time period. Further, this growth in the market was expected to continue for the foreseeable future.

Deposits were critical to Ace Bank's profitability, because they kept the bank's cost of funds low. This was essential if the bank was to be able to make competitively priced loans. Table A4-1 shows the average annual profitability per customer of Ace Bank's consumer deposit products.

The officers were unclear as to why the bank's marketing efforts were not more successful. The bank's main competitors did not seem to be doing anything radically different in their marketing or product offerings. They all agreed, howev-

FIGURE A4-1   PAST AND PROJECTED MARKET SHARE

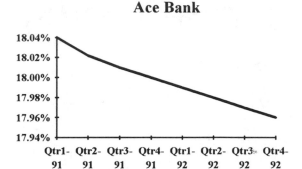

Ace Bank

**TABLE A4-1   AVERAGE ANNUAL PROFITABILITY OF CONSUMER
BANKING PRODUCTS**

Account Type	Average Profit per Customer	Percent of Customers with Account
Checking	$50	90%
Savings	$110	50%
CD/IRA	$300	25%

er, that for the bank's growth to be below the market growth rate was unacceptable, particularly since the bank had sought to increase its market share.

ACE BANK—BACKGROUND

Ace Bank is a subsidiary of a large bank holding company, Time Banc. Ace Bank was formed in August 1990 by the acquisition and merger of two institutions: First Bank and One Bank. Both banks were located in the same market area.

First Bank was a 100-year-old retail (consumer) bank. The bank was known for its conservatism and its "small bank" atmosphere. First Bank's business was primarily conducted inside the branches. The bank had no automatic teller machines (ATMs), although the branches did have drive-through facilities at all locations.

First Bank had 27 branches located throughout its market. Branch managers were promoted from within the company. As a result, all branch managers had been with the bank for five years or longer.

First Bank focused exclusively on consumer business. The largest segment of the bank's customers was the 50 and older age group. Further, the bank's lending was conservative, with a large portion of its loan business being home mortgages and home equity lines of credit.

One Bank was a 12-year-old commercial bank. Its business largely focused on loans to small and middle-market (medium-sized) businesses. While the bank had six branches located throughout its market, the bank's loan officers routinely met with prospects and clients at their homes or offices to conduct business.

Although One Bank was relatively small, it offered its business customers a variety of cash management services. Also, because the majority of its clients were small businesses, a significant percentage of its loan portfolio was asset-based loans.

One Bank's retail banking operations were primarily personal banking relationships with the owners of the businesses with which it was working.

THE MERGER

Time Banc had been looking to establish a presence in First Bank/One Bank's market. In July of 1987, Time Banc acquired One Bank, changing its name to Ace Bank.

While Time Banc wanted Ace Bank to continue its commercial lending to small and middle-market businesses, it also wanted the bank to expand its retail operations. The bank began marketing retail deposit products (checking, savings, etc.) to attract new customers. The marketing efforts largely focused on price, such as free checking or offering slightly higher interest rates than competitors. The bank also purchased eight ATMs, locating one at each of its branches and the remaining two at local shopping centers.

This strategy met with limited success. While the bank did attract new retail customers, Ace Bank still held a relatively small share of the retail banking market. To increase Ace Bank's market share, Time Banc purchased First Bank in August 1990 and merged its operations with Ace Bank. Due to overlap, four First Bank and two Ace Bank branches were shut down. Also, because the merger consolidated operations, 6 percent of the work force of the merged institution was laid off, with the majority being First Bank employees.

The now-larger Ace Bank began to aggressively focus on the retail banking business. As a part of its retail strategy, the bank took the six ATMs located at the old One Bank branches, purchased nine additional ATMs, and located the ATMs at its 15 largest branches. Also, the bank began to heavily promote its checking accounts and its certificates of deposit (CD). Ace Bank chose to sell the large mortgage portfolio it acquired after the purchase of First Bank because it did not provide the return desired by Time Banc. However, the bank continued to offer new mortgages.

The bank also centralized all its commercial lending operations at the main office. This meant that business clients dealt with the branch for their deposits but with the main office for their loans.

Shortly after the merger, the bank noticed a slight decline in its customer base. This had been anticipated and was viewed as an unavoidable consequence of a merger. More disturbing, however, was an increase in the number of customer complaints. While this was at first attributed to the merger, the number of complaints did not diminish over time.

ACE BANK'S CUSTOMERS

The persistence of complaints disturbed the officers of Ace Bank. They decided to conduct research of the bank's depositors to determine their likelihood of leaving the bank and their overall satisfaction with the bank's services. The research also investigated customers' overall satisfaction with six areas of the bank's operations: (1) Tellers, (2) Statements, (3) Interest Rate, (4) Branches, (5) ATMs, and (6) Fees. Further, customers' satisfaction with certain dimensions of the above service areas was examined. For example, customers were asked their satisfaction with the tellers' knowledge, attentiveness, and the length of lines. Table A4-2 displays the areas examined by the research and their corresponding dimensions (when applicable).

The research found that 41 percent of customers felt that Ace Bank's overall level of service exceeded their expectations. This surprised the bank's management, since the research was initiated as a response to customer complaints. However, the research also indicated that 14 percent of customers felt that the bank's overall level of service was much worse than expected. The remaining 46 percent of customers believed that the bank's service overall was about the same as they expected.

TABLE A4-2   AREAS EXAMINED BY MARKET RESEARCH

	Tellers	Statements	Interest	Branches	ATMs	Fees
Dimension 1	Knowledge	Frequency		Hours		
Dimension 2	Attentive	Accuracy		Appearance		
Dimension 3	Lines	Timeliness				
Dimension 4		Understand				

Figure A4-2 shows the percentage of customers who are dissatisfied (expectations were not met) with the bank regarding particular areas of service and their corresponding dimensions. The figure also shows the importance of these various attributes.

The research examined the length of depositors' relationships with Ace Bank (or One Bank/First Bank). For those depositors who were new (beginning their relationship within the past year), the research also sought to determine if they had switched banks or were new to the market. The research indicated that just under 5 percent of the banks' depositors were new to the market. Just over 16 percent of the bank's customers had switched to Ace Bank from another institution within the past year.

CONCLUSION

Ace Bank's management believed that the results of the market research offered insight into the bank's lackluster performance over the past year. The bank's marketing efforts were clearly attracting new customers. Therefore, for the bank's market share to remain relatively flat meant that the bank was not doing a good job of holding on to its customers relative to competitors.

The research also offered insight into what factors cause depositors to defect. The results clearly showed that customers' satisfaction with the bank's tellers was the most important factor in determining their satisfaction with the bank. Further, customers' satisfaction with the tellers' knowledge of the bank's products and policies was the most important element in their overall satisfaction with the tellers. Therefore, the officers decided that the bank must focus on improved training of its front-line personnel.

Ace Bank's management believed that given the relatively high percentage of customers who were delighted (expectations were exceeded) with the bank's service, it was unlikely that the bank could initiate programs that would significantly increase these percentages in a cost-effective manner. They agreed that the best opportunity was in the reduction of dissatisfied (expectations were not met) customers. As a result, training would emphasize methods to prevent and respond to problems.

Currently, the bank spent approximately $30,000 per year on training of tellers to enhance their knowledge of the bank's policies and procedures. Management estimated that a more intensive program, costing around $45,000 per year, could reduce the percentage of customers who were dissatisfied with the tellers' knowledge to 5 percent. They believed it was theoretically possible to reduce the percentage of customers who were dissatisfied with the tellers' knowledge to 0 percent, but that the costs would be prohibitive. If training were dropped entirely, they believed that the percentage of dissatisfied customers would balloon to 50 percent.

FIGURE A4-2   PERCENTAGES OF DISSATISFIED CUSTOMERS AND IMPORTANCE TO RETENTION

PERCENTAGES OF DISSATISFIED CUSTOMERS & IMPORTANCE TO RETENTION
(Management Units 1-5, Dimensions 1-10)

	Tellers		Statements		Interest		Branch		ATMs	
	DIS	IMP	DIS	IMP	DIS	IMP	DIS	IMP	DIS	IMP
Overall	9%	34	18%	16	5%	13	36%	0	36%	0
Dimen 1	14%	72	14%	29			27%	0		
Dimen 2	14%	28	9%	52			23%	100		
Dimen 3	23%	0	27%	19						
Dimen 4			27%	0						

PERCENTAGES OF DISSATISFIED CUSTOMERS & IMPORTANCE TO RETENTION
(Management Units 6-9 + Product, Dimensions 1-10)

	Fees									
	DIS	IMP	DIS	IMP	DIS	IMP	DIS	IMP	DIS	IMP
Overall	36%	13								
Dimen 1										
Dimen 2										
Dimen 3										
Dimen 4										

The question facing Ace Bank's senior officers was whether investing in improved teller training was a good investment. The bank was still profitable and was maintaining its market share. Perhaps cutting costs on things like training would actually improve profits. Whatever they decided, they realized that they would ultimately be held accountable by their parent company, Time Banc.

CASE QUESTIONS

The Ace Bank case is designed so that you can test your abilities using the ROQ software. No answers are provided for this case; however, some questions that should be considered are:

1. Is management correct to focus on the training of its front-line personnel? Why?
2. Approximately what is Ace Bank's current customer retention rate?

Note: The next questions require using the ROQ software.

3. What will be the market share impact of maintaining the current level of teller training (assuming everything else remains the same)? What will be the market share impact if the bank decides to implement the more intensive teller training program?
4. Will the efforts to improve teller training have a positive impact on profits the first year? Why? What about year three? What does this imply about quality improvement programs? (Use a discount rate of 8 percent.)

APPENDIX 5

THE ROQ SOFTWARE USER'S GUIDE

TABLE OF CONTENTS

MINIMUM SYSTEM REQUIREMENTS

- IBM® PC™, AT™, Personal System/2™, or 100% compatible computers
- MS-DOS® 5.0 or higher
- Minimum 2 MB of memory

  Requires a minimum of 490K free conventional memory.

  Requires the use of EMS (Expanded Memory Specification), LIM version 3.2 or later.

  Extended memory is **not** supported.

  Most computer systems are configured to use Extended, not Expanded (EMS), memory. If this is the case for your system, you will need to use a memory manager such as QEMM, 386MAX, BLUEMAX, or the memory manager supplied with MS-DOS 6.0. If Expanded memory is not used, this program may not load or may run extremely slowly. Please consult your DOS manual for more information on expanded memory.

- Graphics adapter (CGA, Hercules, EGA, or VGA)
- 3.5" high-density (1.44 MB) disk drive
- Hard disk drive with a minimum of 2 MB of available space

WHAT IS ROQ SOFTWARE?

The ROQ software is a decision support system designed to identify aspects of a firm's service delivery that have the greatest impact on customer retention. Customers' satisfaction with various elements of a firm's service delivery is linked to their likelihood of continuing to use the firm's services.

Figure A5-1 summarizes how the ROQ software defines the relationship between customer retention and customer satisfaction. Customer retention is viewed as being directly linked to customers' satisfaction with the firm's overall service. Satisfaction with the firm overall is affected by customers' satisfaction with specific business processes (called Management Units in the ROQ software) or the firm's product. Finally, satisfaction with a Management Unit/Product is affected by customers' satisfaction with particular elements of the process (called Dimensions in the ROQ software). For example, a hotel operation may use Room Quality as a Management Unit, with Dimensions being Sufficient Lighting, Bathroom Cleanliness, etc.

The ROQ software allows up to nine Management Units and one Product to be studied. Each Management Unit/Product may contain up to 20 Dimensions.

Customers are presumed to fall into one of three categories regarding their satisfaction with a firm's service: Dissatisfied, Satisfied, or Delighted. A firm's quality-improvement efforts can focus on one of two possible strategies: shifting Dissatisfied customers to Satisfied customers, or shifting Satisfied customers to Delighted customers. These strategies are considered mutually exclusive (meaning shifting Dissatisfied customers to Satisfied and shifting Satisfied customers to Delighted requires different quality-improvement programs).

The ROQ software is divided into two sections: Review and Input. Review screens provide information regarding customers' satisfaction with a firm's service

FIGURE A5-1

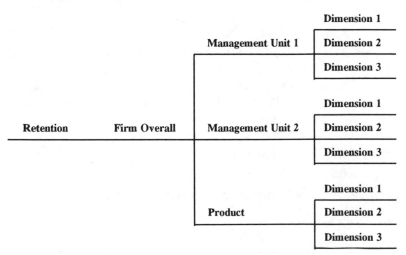

and how satisfaction relates to customer retention. These screens are designed to help you isolate the most promising areas in which to focus the firm's quality-improvement efforts. Input screens allow you to input information regarding the firm's market, costs, etc.

The ROQ software takes into account the costs of achieving a firm's chosen strategy and its corresponding effect on retention to estimate the profit and market share impact of the strategy. Using the ROQ software, a firm can estimate the optimal spending level for a particular quality-improvement effort.

FUNCTION KEYS

The ROQ software uses function keys (F1, F2, F3, etc.) for all screen movement and to access all commands. Below is a list of the function keys used and their purpose.

F1	Help
F2	Forward (move ahead one screen)
F3	Back (move back one screen)
F4	Input (data entry key—used with Input Section only)
F5	Window (switch between review section and Input Section)
F7	Print (print Review Screens or Input Screens)
F10	Quit

INSTALLATION

To run the ROQ software, it must be installed on a hard disk drive. To install the software on your computer's hard disk drive, do the following:

1. Go to the drive to which you wish to install the ROQ software (Example: C:). This is done by typing the letter that corresponds to the desired drive followed by a colon (:) and pressing the [ENTER] key. For example, to go to the C drive, type:

   C: *and press* [ENTER]

Note: Make sure that you are in the root directory (meaning that you are not already in another directory). To get to the root directory, type CD\ and press [ENTER].

2. Create a directory for the software. In this example, we will create a directory named ROQ. This is done by typing:

   MD ROQ *and pressing* [ENTER]

3. Go to the directory that you just created. Using the previous example, this is done by typing:

   CD\ROQ *and pressing* [ENTER]

4. Insert ROQ diskette in drive A (or drive B). Assuming that drive A contains the ROQ software, type:

COPY A:*.* *and press* [ENTER]

The ROQ software should now be installed on your computer's hard drive.

USING THE ROQ SOFTWARE

Getting Started

1. First you should go to the directory that contains the ROQ software. Then, at the DOS prompt, type:

ROQ *and press* [ENTER]

2. Select the type of monitor you are using. The choices are as follows:

Color   To be used with CGA, EGA, or VGA display adapters

LCD/BW   LCD or Black & White monitors using CGA, EGA, or LCD display adapters

Mono/Herc   Monochrome or Hercules display adapters

Note: Some computer systems may use VGA output but have black & white monitors. This may cause the ROQ software to misread your appropriate monitor type, making it difficult if not impossible to read the screen. If this problem exists, before running the ROQ software type MODE BW80 and press [ENTER]

3. Select the type of data you are analyzing. The choices are as follows:

Relationship   Customers maintain long-term relationships

Transaction   Customers conduct periodic transactions

Quit   Return to previous menu

Note: There are only two data sets included with this software. If you are analyzing the Cumberland Hotels data, you will want to select transaction. If you are analyzing the Ace Bank data, you will want to select Relationship. These selections affect how the model calculates market share and revenues.

4. Select the data set you are studying. The choices are:

Hotel   Cumberland Hotels Data

Bank   Ace Bank Data

Quit   Return to Previous Menu

5. Select your appropriate skill level. The choices are as follows:

Novice	Inexperienced ROQ user
Expert	Experienced ROQ user
Quit	Go back to previous menu

Note: If Novice is selected, you will start on an introductory screen, and help screens will automatically appear each time you move to a new screen. If Expert is selected, you will start at the first Review screen, and help screens will appear only if the help key [F1] is pressed.

Review Section

Review screens provide important information regarding customers' satisfaction with a firm's service and how that relates to customer retention. These screens are designed to help you isolate the most promising areas in which to focus the firm's quality-improvement efforts. You move forward and backward through the Review screens by using the F2 and F3 keys, respectively.

The first review screen provides information regarding the percentage of customers who are Dissatisfied, Satisfied, or Delighted with the firm's overall quality of service (see Figure A5-2). This screen shows the maximum percentage of total customers that can be shifted from Dissatisfied to Satisfied or from Satisfied to Delighted regarding the firm's overall service. This information can be used to determine the potential shift available for a particular strategy.

FIGURE A5-2   CURRENT PERCENTAGE OF DISSATISFIED, SATISFIED, AND DELIGHTED CUSTOMERS

CURRENT PERCENTAGE OF DISSATISFIED, SATISFIED, & DELIGHTED CUSTOMERS

Current Percentage of DISSATISFIED Customers Overall	7%
Current Percentage of SATISFIED Customers Overall	87%
Current Percentage of DELIGHTED Customers Overall	6%

* * * POSSIBLE QUALITY IMPROVEMENT STRATEGIES * * *

-	There Are	7%	DISSATISFIED Customers That Can Be Shifted to SATISFIED Customers
-	There Are	87%	SATISFIED Customers That Can Be Shifted to DELIGHTED Customers

**FIGURE A5-3   PERCENTAGES OF DISSATISFIED CUSTOMERS
AND IMPORTANCE TO RETENTION**

	Admit		Physician		Nurses		Inform		Billing	
PERCENTAGES OF DISSATISFIED CUSTOMERS & IMPORTANCE TO RETENTION (Management Units 1-5, Dimensions 1-10)										
	DIS	IMP	DIS	IMP	DIS	IMP	DIS	IMP	DIS	IMP
Overall	12%	15	6%	31	10%	55	9%	43	21%	12
Dimen 1	15%	23	6%	22	8%	65	12%	16	33%	33
Dimen 2	3%	49	12%	35	12%	10	10%	9	21%	9
Dimen 3	7%	37	18%	12	5%	19	6%	63	19%	14
Dimen 4	11%	2	10%	9	9%	5	17%	4		
Dimen 5	0%	0			11%	12	23%	13		
Dimen 6	9%	13								
Dimen 7	13%	61								

The remaining Review screens observe the format shown in shown in Figure A5-3. Each screen will display the percentage of Dissatisfied or Delighted customers for the various Management Units/Product and their related Dimensions. Up to five Management Units/Product and up to 10 Dimensions are displayed per screen.

Each Management Unit/Product will contain two columns. The first column will be marked either DIS (for Dissatisfied) or DEL (for Delighted). This column represents the percentage of total customers that are either Dissatisfied or Delighted regarding a particular aspect of a firm's service. The second column is marked IMP (for Importance). Importance scores can range from a low of 0 to a high of 100. The importance scores for the Overall section describe the strength of the relationship between customers' overall satisfaction with a particular Management Unit/Product and customers' overall satisfaction with a firm's total level of service. The importance scores for the various dimensions describe the strength of the relationship between customers' satisfaction with a particular Dimension and customers' overall satisfaction with its corresponding Management Unit/Product.

Input Section

Input screens allow you to input information regarding a firm's market, costs, etc., in order to evaluate the profit and market share impact of a chosen strategy. To get to the Input section from the Review section, you will press the F5 key. When in the Input section, you will always input data by first pressing the F4 key. You move forward and backward through the Input screens by using the F2 and F3 keys, respectively.

Input Screen 1

Figure A5-4 shows the first Input screen. This screen asks you to provide information regarding your market. You are asked to provide the following: (1) Market Size, (2) Expected Growth Rate, (3) Transactions, and (4) Average Profit.

MARKET SIZE:	The current size of your market, in thousands (Example: input 100 for 100,000).
GROWTH RATE:	The rate per period at which your market is expanding or contracting (Example: input 10 for 10 percent).
TRANSACTIONS:	The number of transactions the average customer makes per period
AVERAGE PROFIT:	The profit impact of a single transaction or customer on average in thousands of dollars (Example: input .1 for $100).

Input Screen 2

Figure A5-5 shows the second Input screen. In this screen you are asked to identify the area in which you are going to focus your quality-improvement efforts. You are given the option of focusing your efforts on improving customers' satisfaction with the firm overall, a particular Management Unit/Product overall, or a specific

FIGURE A5-4   CUSTOMER AND MARKET INFORMATION

CUSTOMER AND MARKET INFORMATION	
Current Number of Transactions in the Market (000)	180
Market Growth Rate per Period (Example: 5 for 5%)	2.5%
Average Number of Transactions per Customer (MAX = 50)	2.4
Average Dollar Profit per Customer per Transaction ($000)	.21

FIGURE A5-5   ATTRIBUTE UNDER CONSIDERATION

ATTRIBUTE UNDER CONSIDERATION

Firm Overall (0), Mgmt. Unit (1), or Product (2)?	1
Which Managerial Unit (1-9)?	1
Unit Overall (1) or a Specific Dimension (2)?	2
Which dimension (1-20)?	7

Admit
Dim 7

Dimension. After pressing the F4 key, you will be prompted with several questions designed to isolate the area under consideration. The process is as follows:

1. You will be asked if you want to focus your efforts on the Firm Overall or on a particular Managerial Unit/Product.
2. If a Management Unit is selected, you will be asked to specify which unit.
3. If a Management Unit or Product has been selected, you will be asked whether your efforts will focus on the Unit or Product overall or on a specific Dimension within the category.
4. If a specific Dimension has been selected, you will be asked to provide the number that corresponds to that particular dimension.

After the data have been entered, the double-lined box in the lower-right corner of the screen will display the area you have selected.

Input Screens 3 and 4

Figure A5-6 displays the next Input screens. You are asked to provide the percentage of DISSATISFIED or DELIGHTED customers you would expect given various levels of spending to improve service for the area under consideration. You must choose whether to focus your efforts on reducing DISSATISFIED customers or on increasing DELIGHTED customers. Spending Level 1 must be less than Spending Level 2. The model assumes that greater spending will reduce the percentage of DISSATISFIED customers or increase the percentage of DELIGHTED customers.

This information will be used to estimate the likely percentages of DISSATISFIED or DELIGHTED customers at various spending levels. This function will then be

FIGURE A5-6   EXPECTED PERCENTAGE OF DISSATISFIED CUSTOMERS

EXPECTED PERCENTAGE OF DISSATISFIED CUSTOMERS

	$ Level	% DIS
Minimum Spending Level	$ 0	20.00%
Spending Level 1 ($000)	7	12.50%
Spending Level 2 ($000)	15	5.00%
Maximum Spending Level	$ Infinity	1.00%

12.50% are DISSATISFIED and   6.00% are DELIGHTED regarding

Admit	Dim 1

PERCENT DISSATISFIED AND CORRESPONDING RETENTION RATE

$000	% Dissat	Retention
$0.00	20.00%	91.64%
$2.00	19.33%	91.66%
$4.00	17.09%	91.73%
$6.00	14.03%	91.83%
$8.00	11.07%	91.99%
$10.00	8.67%	92.04%
$12.00	6.85%	92.07%
$14.00	5.52%	92.09%
$16.00	4.55%	92.11%
$18.00	3.84%	92.12%
$20.00	3.30%	92.13%
$22.00	2.89%	92.14%
$24.00	2.58%	92.14%
$26.00	2.33%	92.14%
$28.00	2.14%	92.15%
$30.00	1.98%	92.15%

graphed for you over whatever spending range you believe is appropriate; you will be asked to provide the MINIMUM and MAXIMUM spending levels you wish to consider. You will also be given the opportunity to save the graph for printing later.

Input Screens 5 and 6

Figure A5-7 shows Input screens 5 and 6. These screens are used to estimate the profit impact to the firm given various spending levels to improve service. The first question asks for any cost savings that you expect to achieve as a result of your program. You must assign when these cost savings are expected to begin and end with periods ranging from 1 to 10 (where 1 is the first period following the initiation of the program, and 10 is 10 periods after the start of the program).

Likewise, the expenses of your firm's service quality program can be assigned from 0 to 10 (where 0 is the current period, 1 is the first period following the initiation of the program, and 10 is 10 periods after the start of the program).

You are also asked to provide the number of periods you want to consider when calculating the net present value of the effort, as well as the interest rate (discount rate) desired.

This function will be graphed for you over whatever spending range you believe is appropriate; you will be asked to provide the MINIMUM and MAXIMUM spending levels you wish to consider. You will also be given the opportunity to save the graph for printing later.

Input Screens 7 and 8

Figure A5-8 shows the final two input screens in the ROQ model. These screens show the firm's expected Market Share over the 10 periods following the start of the quality-improvement program given specified spending levels to improve service. You can provide up to four possible spending levels you wish to consider.

Simply input the number of spending options that you want to consider and then the appropriate dollar amounts for each option. Spending levels should be in thousands of dollars (Example: 5 for $5,000). Your firm's expected market share for each of the spending levels will then be graphed over 10 periods. You will also be given the opportunity to save the graph for printing later.

PRINTING THE INPUT AND REVIEW SCREENS

The ROQ software allows you to print the Review and Input screens on Hewlett Packard laser printers. The following printers are supported:

HP LaserJet+

HP LaserJet II

HP LaserJet IIP

HP LaserJet IID

HP LaserJet III

FIGURE A5-7   PROFIT IMPACT OF SERVICE QUALITY IMPROVEMENT EFFORT

PROFIT IMPACT OF SERVICE QUALITY IMPROVEMENT EFFORT

Expected Annual Cost Savings from Effort ($000)?	0
First Period for Cost Savings (1-10)?	1
Last Period for Cost Savings (1-10)?	10
First Period for Expenses (1-10)?	0
Last Period for Expenses (0-10)?	10
Number of Periods for NPV Calculation (1-10)?	3
Interest Rate (Example: 10 for 10%)?	9.00%

NPV OF IMPROVEMENT EFFORT AT DIFFERENT SPENDING LEVELS ($000)

FOCUS		$ LEVEL	NPV	COST SAVINGS	
FOCUS		$0.00	$495	COST SAVINGS	
		$2.00	$497		
Admit		$4.00	$509	$ Amount	0
		$6.00	$527		
Dim 1		$8.00	$545	1st Period	1
		$10.00	$555		
Dissatisfied		$12.00	$560	Last Period	10
		$14.00	$561		
		$16.00	$561		
		$18.00	$559		
NPV		$20.00	$555	EXPENSES	
		$22.00	$551		
Periods	3	$24.00	$546	1st Period	0
		$26.00	$541		
Int. Rate	9.00%	$28.00	$536	Last Period	10
		$30.00	$530		

FIGURE A5-8   MARKET SHARE IMPACT OF SERVICE QUALITY IMPROVEMENT EFFORT

MARKET SHARE IMPACT OF SERVICE QUALITY IMPROVEMENT EFFORT

Number of Spending Options (1-4)?	4
Spending Level 1 ($000)	0
Spending Level 2 ($000)	5
Spending Level 3 ($000)	10
Spending Level 4 ($000)	30

Firm   Overall   Delighted

MARKET SHARE IMPACT OF SERVICE QUALITY IMPROVEMENT EFFORT

	Level 1	Level 2	Level 3	Level 4
Current	13.00%	13.00%	13.00%	13.00%
Period 1	12.98%	13.03%	13.04%	13.05%
Period 2	12.97%	13.06%	13.08%	13.08%
Period 3	12.95%	13.08%	13.11%	13.11%
Period 4	12.94%	13.10%	13.13%	13.14%
Period 5	12.93%	13.12%	13.15%	13.16%
Period 6	12.93%	13.13%	13.17%	13.18%
Period 7	12.92%	13.14%	13.18%	13.19%
Period 8	12.92%	13.15%	13.19%	13.21%
Period 9	12.91%	13.16%	13.20%	13.22%
Period 10	12.91%	13.16%	13.21%	13.22%
($000)	0	5	10	30

HP LaserJet IIIP

HP LaserJet IIID

HP LaserJet IIISI

Laser printers that are 100 percent compatible with the above printers are also supported. To print, simply press the F7 key and select either Review or Input screens.

USING HELP

Each screen has a corresponding help screen.

QUITTING THE ROQ SOFTWARE

To leave the ROQ software, press the F10 key. You will be given the opportunity to save your work prior to leaving.

PRINTING ROQ GRAPHS[1]

Printing graphs is done outside of the ROQ software with a program called PicPrint. The PicPrint program prints graphs (that you saved as .PIC files in the ROQ software) on a printer or plotter. PicPrint includes a number of different options that let you control the way in which graphs are printed. For example, PicPrint lets you specify where your graph will use colors, what character styles you will employ, how the output will be positioned on the page, and what type of printer (or plotter) you are using.

Note: PicPrint requires about 160 KB of free memory to operate. If you print full-page graphs at high resolution, then PicPrint may require more memory.

STARTING PICPRINT

To use PicPrint, you must have saved the graph while running the ROQ software. You run PicPrint from DOS by typing:

PICPRT *and pressing* [ENTER]

You can also add the graph file name to the command line to have PicPrt print the graph immediately. For example:

PICPRT mygraph.pic

USING PICPRINT

When you load PicPrint for the first time, your screen will look like Figure A5-9. At the top of the screen, you see a command menu. The PicPrint program contains the following options:

Select   Lets you choose the graph (.PIC) files you want to print. You can also preview on the screen what the printout of the graph will look like.

FIGURE A5-9

PicPrint 1.4   Copyright 1990-1992 Baler Software Corporation All Rights Reserved

Select Options Go Configure Align Page Quit
Select pictures

SELECTIONS	OPTIONS			CONFIGURATION
SELECTED GRAPHS	COLORS	SIZE HALF		PICTURE DIRECTORY
	Grid: Black	Left Margin:	.750	
	A Range: Black	Top Margin:	.395	
	B Range: Black	Width:	6.500	GRAPHIC DEVICE
	C Range: Black	Height:	4.691	
	D Range: Black	Rotation:	.000	HP LaserJet 300dpi
	E Range: Black			Parallel
	F Range: Black	MODES		
				VIDEO DEVICE
	FONTS	Eject: No		
		Pause: No		IBM VGA/EGA
	1: Roman Medium			
	2: Sans Serif Light			PAGE
				Length 18.000
				Width 8.000

Options	Lets you choose from the available options, such as color, font, and picture size.
Go	Starts printing the graphs you indicated at the graphics device you specified.
Configure	Lets you specify the printer, type of video adapter, graphs directory, and other hardware settings for your computer.
Align	Sets the current paper position in the printer as the top of the page.
Page	Advances the paper in the printer to the top of the next page.
Quit	Returns you to the DOS command Prompt.

PicPrint displays the current configuration settings in the area of the screen below the main menu. This area of the screen holds the following information:

SELECTED GRAPHS	MODES
COLORS	PICTURE DIRECTORY
FONTS	GRAPHICS DEVICE

SIZE   VIDEO DEVICE

MANUAL   PAGE SIZE

If this is the first time you have started the PicPrint program, you can interpret the default settings in this manner. There are no graph files SELECTED for printing. Because the default color option isn't on, the grid will be printed in black, and each of the data ranges will be black. If your graph has a title, the first line will be printed in Font 1 (Roman Medium); any other text on your graph, such as the second line of the title, legends, and axes titles, will be printed in Font 2 (Italic Medium).

The SIZE settings are measured in inches from the edge of the paper you are using. For example, the left margin is .75 inches from the left edge of the paper. The special mode settings (eject and page) are both set No. In other words, PicPrint will not pause and will not eject a blank page of paper between each selected graph print job.

The default picture directory is set to your current directory. This is where PicPrint will look for graph (.PIC) files. The default graphics printing device is an HP LaserJet. The default video device (the type of monitor you have) is an IBM EGA/VGA. The default page size is the standard 8 1/2 by 11 inches.

CHANGING THE CONFIGURATION SETTINGS

Before you print a graph, check that all the PicPrint configuration settings are correct. To change PicPrint's default configuration settings, choose Configure at the main menu. This displays the following options:

Files

Device

Page

Interface

Video

Save

Reset

Quit

Files

The Configure Files command lets you indicate in which drive and directory PicPrint defaults to look for graph (.PIC) files. When you select Configure Files, you see the prompt:

Enter where to search for picture files

On the line beneath the prompt, you see the blinking cursor. The default drive and directory or the last drive you specified are displayed here. To change the drive and/or directory, type the appropriate path, then press [ENTER]. You return to the Configure menu, and PicPrint displays the current directory.

Note: When you enter a new picture directory, it will be active only for the current PicPrint session. If you want PicPrint to use the new picture directory as the default, you must issue the Configure Save command.

Device

The Configure Device command lets you select from different graphics devices (graphics printers and plotters). To select a device, you must first highlight the graphics device of your choice, then press the space bar to mark it. The keys you use to make your selection are listed to the right of the Configure Device screen. Select your graphics device from the list. (You can press the HOME key to get to the first option in the list and END to see the last option in the list.)

Press the up or down arrow keys to move the highlighted menu pointer to the graphics device you want, then press the space bar to mark your choice. You will see a # (pound sign) to the left of the marked choice. Press [ENTER] to return to the Configure menu. If you want to change the current selection, move the menu pointer to another device option and press the space bar. You can have only one graphics device selected at any time.

Note: When you mark a new graphics device, it will be active only for the current PicPrint session. If you want PicPrint to use the new graphics device selection as the default, you must issue the Configure Save command.

Page

The Configure Page command lets you change the length and width of your paper. The page length and width is measured in inches; the default length is 11 inches, and the default width is 8 1/2 inches.

When you select the Configure Page command, you will see the following options:

Length   Width   Quit

To change the page length, select Configure Page Length. You will see the following prompt:

Enter page length in inches

Beneath the prompt, you will see the blinking cursor following the decimal value of the default page-length setting. Type the value you want, and press [ENTER]. The new page length will be active for the current PicPrint session. Select Configure Save to update the PicPrint configuration file.

To change the page width, select Configure Page Width. You will see the following prompt:

Enter page width in inches

Beneath the prompt, you will see the blinking cursor following the decimal value of the default page-width setting. Type the value you want, and press [ENTER]. The new page width will be active for the current PicPrint session.

Interface

The Configure Interface option tells PicPrint where to send the .PIC file; the following choices are available:

    1 - The first parallel port (LPT1).
    2 - The first serial port (COM1).
    3 - The second parallel port (LPT2)
    4 - The second serial port (COM2).
    5 - A disk file (SPOOL.FIL).

If you have a parallel printer attached to the first parallel device port (LPT1), choose option 1. If you have a serial printer attached to the first serial port (COM1), choose option 2. If you have a parallel printer attached to the second parallel device port (LPT 2), choose option 3. If you have a serial printer attached to the second serial port (COM 2), choose option 4. Choose option 5 when you want PicPrint to create a disk file, called SPOOL.FIL, to be used with a print spooler.

Note: Use DOS COPY / B (binary) command when you are ready to print SPOOL.FIL.

If you choose either of the serial port options (2 or 4), you will have to select from nine baud rates. A baud rate is the speed (measured in characters per second) at which the serial device will transmit characters. The following choices are available:

    1 - 110 baud
    2 - 150 baud
    3 - 300 baud
    4 - 600 baud
    5 - 1200 baud
    6 - 2300 baud
    7 - 4800 baud
    8 - 9600 baud
    9 - 19200 baud

If you are unsure of the optimum baud rate for your printer (or plotter), refer to your printer (or plotter) manual.

Video

The video device is the type of monitor you have. PicPrint needs to know what type of video device you are using so you can display (or preview) on your screen what the printout of your graph will look like.

To preview a graph, you choose Select from PicPrint's Main Menu, highlight a graph of your choice, then press F10 (Graph) key. You will see a picture of your graph, with the fonts (and colors, if your monitor allows them) you selected with the Options command. Press [ESC] to return to the Select menu.

Note: The preview of your graph will not look exactly like the printout. This is because the screen image and the printed image have different proportions. Your graphic images will be printed correctly.

The Configure Video command lets you choose from four video display adapter options:

Monochrome

CGA

VGA/EGA

Hercules

If you do not know what type of monitor adapter your computer uses, read your computer's hardware manual. Select Monochrome if your computer has no graphic display capability. Select CGA if your computer has an IBM Color Graphics Adapter (or compatible). Select VGA/EGA if your computer has an IBM Enhanced Graphics Adapter or an IBM Video Graphics Array adapter (or compatible). Select Hercules if your computer has a Hercules monochrome adapter card.

Save

When you select Configure Save, PicPrint stores all the current settings on disk. The next time you start the program, these are the default settings.

Reset

The Configure Reset command lets you reset all the settings you have changed in the current PicPrint session back to the most recently saved settings. When you select Reset, PicPrint gives you the opportunity to cancel the command if you wish. Select Yes to reset the current values of the last stored configuration settings; select No to cancel the command and return to the Configure menu.

Quit

Choose the Quit option from the Configure menu to return to the PicPrint Main Menu.

PRINTING GRAPHS

All the other PicPrint options, excluding Configure, deal with printing graphs. You need to select the graph(s) you want to print and choose new options (if any) that you want to print with, including colors, fonts, size, and positioning of the graph.

Select

Choose the Select option to mark which graph (.PIC) files in the current picture directory you want to print. You can select one or more files. When you choose the Select option, you will see a menu listing all the .PIC files in the current picture directory. If you do not see any files listed, check that you specified the correct directory and drive with the Configure Files Pictures command.

The Select command lets you choose from a menu listing all the .PIC files in the current default directory. The keys you use to make your selection are listed at the right of the Select screen. To select a file, you must first highlight the file of your choice, then mark it and press [ENTER]. If you want to print more than one graph in this PicPrint session, mark the graphs in the order you want them printed, then press [ENTER]. When you return to the Main Menu, you will see each graph you selected in the order in which you marked them.

Press the arrow keys to move the highlighted menu pointer to the file you want, then press the space bar to mark your choice. You will see a # to the left of the marked choice. If you want to "de-select" a file selection, move the menu pointer to that file name and press the space bar so that the # disappears. Press [ENTER] when you have marked all the graphs you intend to print, PicPrint returns you to the Main Menu. Once you press [ENTER], you will see the selected file names listed in the PicPrint display in the order in which you marked them.

You can also preview a graph on the screen to see how it will look when it's printed. To preview a graph, press the F10 (Graph) key when the menu pointer is highlighting a graph in the Select menu. It is not necessary to mark a graph to review it.

Options

The Options command lets you indicate what characteristics you want your graph to be printed with. The following Options are available:

Color

Size

Pause

Eject

Quit

Font

Color

When you select Option Color, you are assigning color to different ranges you have specified in your graph. You will be allowed to change colors only if you have selected a graphics device (with PicPrint's Configure Device command) that supports multiple colors. You can assign colors to the following ranges:

Grid   The grid is the horizontal and/or vertical lines assigned to your graph. Also included in the Grid setting are both the graph's and the axes' titles.

A   The worksheet data defined for the A data range of your graph and the corresponding legend for the A data range.

B   The worksheet data defined for the B data range of your graph and the corresponding legend for the B data range.

C   The worksheet data defined for the C data range of your graph and the corresponding legend for the C data range.

D   The worksheet data defined for the D data range of your graph and the corresponding legend for the D data range.

E   The worksheet data defined for the E data range of your graph and the corresponding legend for the E data range.

F   The worksheet data defined for the F data range of your graph and the corresponding legend for the F data range.

After you select which data ranges you want to print in color, PicPrint updates the display to reflect your changes.

Font

Options Font lets you select the primary (1) and alternate (2) fonts that you want for the text portions of your graph. The first line of the graph's title prints in the primary font, and any other text (such as legends, axes titles, and so on) prints in the secondary font.

The following fonts are available for the primary and second fonts:

Gothic Bold	Roman Medium (Small)
Italic Bold	Sans Serif Light
Italic Medium	Sans Serif Medium
Italic Medium (Small)	Script Light
Medieval Bold	Script Medium
Roman Bold	Teutonic Bold
Roman Medium	

You select a font in the same manner that you select a file (Select File) or a device (Configure Device). Press the highlighted menu pointer to select the font you want, then press the space bar to mark your choice. You will see a # to the left of the marked choice. If you want to "de-select" a font selection, move the menu pointer to that file name and press the space bar so that the # disappears. Press [ENTER] after you mark your choice, and PicPrint returns you to the Configure menu. Once you press [ENTER], you will see the selected fonts listed in the PicPrint display.

Size

The Size option lets you specify how you want the graph positioned on the paper. When you select Size, you see the following menu options:

Full

Half

Manual

Quit

Selecting the Full option prints the graph full size, taking up an entire 8 1/2 x 11-inch page. When you select Option Size Full, PicPrint automatically rotates your graph 90 degrees so that your graph fills the page.

When you select the Half Option, the printed graph occupies only one-half of the 8 1/2 x 11-inch page.

When you select Manual, you have the following menu choices:

Left   Top   Width   Height   Rotation   Quit

The Left setting specifies the size of the left margin in inches. The Top setting specifies the number of inches for the top margin. The Width option indicates how wide the printed graph will be; Height indicates how many inches the vertical length of the graph will be. You can change the Width and Height setting to be bigger than the width and height of the page for multipage graphs.

Rotation adjusts the rotation of the graph on the page in a counterclockwise direction. For example, if you have Rotation set to .000, your graph will be printed with its x-axis parallel to the width of the page. If you have 90 degrees of Rotation, the X-axis will run along the height of the paper (where the y-axis usually is) perpendicular to the page width.

Pause

When you select the Pause option, PicPrint tells the printer to pause after printing each graph, or each page of a multiple-page print job.

Eject

When you select the Eject option, PicPrint tells the printer to eject a blank page after printing each individual graph, or between printing each page of a multiple-page graph.

Note: If you are printing with a high-resolution printer (like a laser printer set to 300 dpi), PicPrint may not be able to complete the rendering of the graph in one pass. When this happens, PicPrint will print parts of your graph across many pages. To print a high-resolution graph on one page, set Eject to No.

Quit

When you select Quit from the Options menu, PicPrint returns to the Main Menu.

Go

Select Go when you have set all the configuration settings correctly, and you have selected the graph (.PIC) files that you want to print in the PicPrint session. Depending on your computer and the type of printer you are using, graphs should start printing in just a few seconds.

As PicPrint generates a graph and sends it to your printer or plotter, you will see a message similar to the following:

Generating Picture B:\LINEGRAF.PIC

When all the graphs have printed, PicPrint returns to the Main Menu.

Align

The Align option lets you tell your graphics device where to set the top of the page (sometimes called the top of the form). When you select Align, PicPrint tells your printer that it's currently at the top of the page.

Page

The Page option tells your printer to advance the paper to the top of the next page (you can define the top of the page with the Align option).

Quit

Select the Quit option when you are ready to end the current PicPrint session. When you specify Quit, PicPrint lets you choose No and return to the Main Menu, or Yes to exit PicPrint and return to the DOS system prompt.

NOTE

1. This section was adapted from *Baler Users Guide,* © 1992 Baler Software Corporation. Used with permission.

INDEX

ABOUT THE AUTHORS

Roland T. Rust is Professor and Head of the Marketing area at the Owen Graduate School of Management at Vanderbilt University, where he teaches courses in service quality, customer service, and customer satisfaction. As Director of the Owen School's Center for Services Marketing, he works with leading businesses and non-profit organizations to establish and evaluate their service quality and customer satisfaction programs.

He has consulted, conducted training programs, and supervised research studies for such companies as AT&T, Procter & Gamble, Promus Companies, Humana, HCA, Northern Telecom, Marathon Oil, American Airlines, Consumers Union, Sierra Club, the Concord Group, and NationsBank. Professor Rust is the author of numerous publications in Marketing, Advertising, and related fields, and he serves on the editorial review boards of six journals including the *Journal of Marketing Research, Marketing Science,* and the *Journal of Retailing.* He is the Co-Chair of the "Frontiers in Services" conference, co-sponsored by the American Marketing Association and Vanderbilt's Center for Services Marketing.

Professor Rust is a member of the Services Marketing Council of the American Marketing Association. His current research centers on the financial impact of quality improvement programs. He holds MBA and Ph.D. degrees from the University of North Carolina at Chapel Hill.

Anthony J. Zahorik is Assistant Professor of Marketing at the Owen Graduate School of Management at Vanderbilt University. He previously served as Associate Dean of the Owen School. His research includes studies of the financial impact of service quality improvements in a wide range of service industries. He has also consulted and taught executive seminars on service quality programs and marketing planning to firms in many industries, including health care and banking. Zahorik holds an M.A. in mathematics from the University of Illinois. He has also earned MBA and Ph.D. degrees from Cornell University.

Timothy L. Keiningham is the Director of Research Services for the National Association of Bank Directors. His work focuses on issues regarding the statutory and fiduciary responsibility for boards of directors of financial institutions. Keiningham is also Product Development Manager for FISI-Madison Financial, a bank marketing firm. His work for FISI-Madison Financial focuses on creating products/services for retail customers of financial institutions and on studying aspects of service delivery that affect customers' attraction to financial institutions and their products.

ORDER YOUR ROQ DEMONSTRATION DISKETTE!

The ROQ demonstration diskette provides a tangible representation of how an ROQ system should work. The software lets you practice the use of the ROQ model with two sample data sets and is designed to accompany the appendices in this book.

If you wish to receive the ROQ demonstration diskette discussed in this book, please call our Customer Service Center at:

1-800-776-2871

Or write to:

Probus Publishing Company
1925 North Clybourn Avenue
Chicago, IL 60614

Ask for the ROQ MODEL DISKETTE, Order Number 547D

THE PRICE IS $10.00 FOR EACH COPY OF THE MODEL, PLUS A $5.50 SHIPPING AND HANDLING FEE. ORDERS MUST BE PREPAID BY CREDIT CARD OR CHECK PRIOR TO SHIPMENT.

This offer is subject to availability and may be canceled by the Publisher at any time.